Measurement: A Very Short Introduction

VERY SHORT INTRODUCTIONS are for anyone wanting a stimulating and accessible way into a new subject. They are written by experts, and have been translated into more than 40 different languages.

The series began in 1995, and now covers a wide variety of topics in every discipline. The VSI library now contains over 500 volumes—a Very Short Introduction to everything from Psychology and Philosophy of Science to American History and Relativity—and continues to grow in every subject area.

Very Short Introductions available now:

ACCOUNTING Christopher Nobes
ADOLESCENCE Peter K. Smith
ADVERTISING Winston Fletcher
AFRICAN AMERICAN RELIGION Eddie S. Glaude Jr
AFRICAN HISTORY John Parker and Richard Rathbone
AFRICAN RELIGIONS Jacob K. Olupona
AGNOSTICISM Robin Le Poidevin
AGRICULTURE Paul Brassley and Richard Soffe
ALEXANDER THE GREAT Hugh Bowden
ALGEBRA Peter M. Higgins
AMERICAN HISTORY Paul S. Boyer
AMERICAN IMMIGRATION David A. Gerber
AMERICAN LEGAL HISTORY G. Edward White
AMERICAN POLITICAL HISTORY Donald Critchlow
AMERICAN POLITICAL PARTIES AND ELECTIONS L. Sandy Maisel
AMERICAN POLITICS Richard M. Valelly
THE AMERICAN PRESIDENCY Charles O. Jones
THE AMERICAN REVOLUTION Robert J. Allison
AMERICAN SLAVERY Heather Andrea Williams
THE AMERICAN WEST Stephen Aron
AMERICAN WOMEN'S HISTORY Susan Ware
ANAESTHESIA Aidan O'Donnell
ANARCHISM Colin Ward
ANCIENT ASSYRIA Karen Radner
ANCIENT EGYPT Ian Shaw
ANCIENT EGYPTIAN ART AND ARCHITECTURE Christina Riggs
ANCIENT GREECE Paul Cartledge
THE ANCIENT NEAR EAST Amanda H. Podany
ANCIENT PHILOSOPHY Julia Annas
ANCIENT WARFARE Harry Sidebottom
ANGELS David Albert Jones
ANGLICANISM Mark Chapman
THE ANGLO-SAXON AGE John Blair
THE ANIMAL KINGDOM Peter Holland
ANIMAL RIGHTS David DeGrazia
THE ANTARCTIC Klaus Dodds
ANTISEMITISM Steven Beller
ANXIETY Daniel Freeman and Jason Freeman
THE APOCRYPHAL GOSPELS Paul Foster
ARCHAEOLOGY Paul Bahn
ARCHITECTURE Andrew Ballantyne
ARISTOCRACY William Doyle
ARISTOTLE Jonathan Barnes
ART HISTORY Dana Arnold
ART THEORY Cynthia Freeland
ASTROBIOLOGY David C. Catling
ASTROPHYSICS James Binney
ATHEISM Julian Baggini
AUGUSTINE Henry Chadwick
AUSTRALIA Kenneth Morgan
AUTISM Uta Frith
THE AVANT GARDE David Cottington

THE AZTECS David Carrasco
BABYLONIA Trevor Bryce
BACTERIA Sebastian G. B. Amyes
BARTHES Jonathan Culler
THE BEATS David Sterritt
BEAUTY Roger Scruton
BESTSELLERS John Sutherland
THE BIBLE John Riches
BIBLICAL ARCHAEOLOGY Eric H. Cline
BIOGRAPHY Hermione Lee
BLACK HOLES Katherine Blundell
BLOOD Chris Cooper
THE BLUES Elijah Wald
THE BODY Chris Shilling
THE BOOK OF MORMON Terryl Givens
BORDERS Alexander C. Diener and Joshua Hagen
THE BRAIN Michael O'Shea
THE BRICS Andrew F. Cooper
THE BRITISH CONSTITUTION Martin Loughlin
THE BRITISH EMPIRE Ashley Jackson
BRITISH POLITICS Anthony Wright
BUDDHA Michael Carrithers
BUDDHISM Damien Keown
BUDDHIST ETHICS Damien Keown
BYZANTIUM Peter Sarris
CANCER Nicholas James
CAPITALISM James Fulcher
CATHOLICISM Gerald O'Collins
CAUSATION Stephen Mumford and Rani Lill Anjum
THE CELL Terence Allen and Graham Cowling
THE CELTS Barry Cunliffe
CHAOS Leonard Smith
CHEMISTRY Peter Atkins
CHILD PSYCHOLOGY Usha Goswami
CHILDREN'S LITERATURE Kimberley Reynolds
CHINESE LITERATURE Sabina Knight
CHOICE THEORY Michael Allingham
CHRISTIAN ART Beth Williamson
CHRISTIAN ETHICS D. Stephen Long
CHRISTIANITY Linda Woodhead
CITIZENSHIP Richard Bellamy
CIVIL ENGINEERING David Muir Wood
CLASSICAL LITERATURE William Allan
CLASSICAL MYTHOLOGY Helen Morales
CLASSICS Mary Beard and John Henderson
CLAUSEWITZ Michael Howard
CLIMATE Mark Maslin
CLIMATE CHANGE Mark Maslin
COGNITIVE NEUROSCIENCE Richard Passingham
THE COLD WAR Robert McMahon
COLONIAL AMERICA Alan Taylor
COLONIAL LATIN AMERICAN LITERATURE Rolena Adorno
COMBINATORICS Robin Wilson
COMEDY Matthew Bevis
COMMUNISM Leslie Holmes
COMPLEXITY John H. Holland
THE COMPUTER Darrel Ince
COMPUTER SCIENCE Subrata Dasgupta
CONFUCIANISM Daniel K. Gardner
THE CONQUISTADORS Matthew Restall and Felipe Fernández-Armesto
CONSCIENCE Paul Strohm
CONSCIOUSNESS Susan Blackmore
CONTEMPORARY ART Julian Stallabrass
CONTEMPORARY FICTION Robert Eaglestone
CONTINENTAL PHILOSOPHY Simon Critchley
COPERNICUS Owen Gingerich
CORAL REEFS Charles Sheppard
CORPORATE SOCIAL RESPONSIBILITY Jeremy Moon
CORRUPTION Leslie Holmes
COSMOLOGY Peter Coles
CRIME FICTION Richard Bradford
CRIMINAL JUSTICE Julian V. Roberts
CRITICAL THEORY Stephen Eric Bronner
THE CRUSADES Christopher Tyerman
CRYPTOGRAPHY Fred Piper and Sean Murphy
CRYSTALLOGRAPHY A. M. Glazer
THE CULTURAL REVOLUTION Richard Curt Kraus
DADA AND SURREALISM David Hopkins
DANTE Peter Hainsworth and David Robey
DARWIN Jonathan Howard
THE DEAD SEA SCROLLS Timothy Lim
DECOLONIZATION Dane Kennedy
DEMOCRACY Bernard Crick

DERRIDA Simon Glendinning
DESCARTES Tom Sorell
DESERTS Nick Middleton
DESIGN John Heskett
DEVELOPMENTAL BIOLOGY Lewis Wolpert
THE DEVIL Darren Oldridge
DIASPORA Kevin Kenny
DICTIONARIES Lynda Mugglestone
DINOSAURS David Norman
DIPLOMACY Joseph M. Siracusa
DOCUMENTARY FILM Patricia Aufderheide
DREAMING J. Allan Hobson
DRUGS Les Iversen
DRUIDS Barry Cunliffe
EARLY MUSIC Thomas Forrest Kelly
THE EARTH Martin Redfern
EARTH SYSTEM SCIENCE Tim Lenton
ECONOMICS Partha Dasgupta
EDUCATION Gary Thomas
EGYPTIAN MYTH Geraldine Pinch
EIGHTEENTH-CENTURY BRITAIN Paul Langford
THE ELEMENTS Philip Ball
EMOTION Dylan Evans
EMPIRE Stephen Howe
ENGELS Terrell Carver
ENGINEERING David Blockley
ENGLISH LITERATURE Jonathan Bate
THE ENLIGHTENMENT John Robertson
ENTREPRENEURSHIP Paul Westhead and Mike Wright
ENVIRONMENTAL ECONOMICS Stephen Smith
ENVIRONMENTAL POLITICS Andrew Dobson
EPICUREANISM Catherine Wilson
EPIDEMIOLOGY Rodolfo Saracci
ETHICS Simon Blackburn
ETHNOMUSICOLOGY Timothy Rice
THE ETRUSCANS Christopher Smith
EUGENICS Philippa Levine
THE EUROPEAN UNION John Pinder and Simon Usherwood
EVOLUTION Brian and Deborah Charlesworth
EXISTENTIALISM Thomas Flynn
EXPLORATION Stewart A. Weaver
THE EYE Michael Land
FAMILY LAW Jonathan Herring
FASCISM Kevin Passmore
FASHION Rebecca Arnold
FEMINISM Margaret Walters
FILM Michael Wood
FILM MUSIC Kathryn Kalinak
THE FIRST WORLD WAR Michael Howard
FOLK MUSIC Mark Slobin
FOOD John Krebs
FORENSIC PSYCHOLOGY David Canter
FORENSIC SCIENCE Jim Fraser
FORESTS Jaboury Ghazoul
FOSSILS Keith Thomson
FOUCAULT Gary Gutting
THE FOUNDING FATHERS R. B. Bernstein
FRACTALS Kenneth Falconer
FREE SPEECH Nigel Warburton
FREE WILL Thomas Pink
FRENCH LITERATURE John D. Lyons
THE FRENCH REVOLUTION William Doyle
FREUD Anthony Storr
FUNDAMENTALISM Malise Ruthven
FUNGI Nicholas P. Money
GALAXIES John Gribbin
GALILEO Stillman Drake
GAME THEORY Ken Binmore
GANDHI Bhikhu Parekh
GENES Jonathan Slack
GENIUS Andrew Robinson
GEOGRAPHY John Matthews and David Herbert
GEOPOLITICS Klaus Dodds
GERMAN LITERATURE Nicholas Boyle
GERMAN PHILOSOPHY Andrew Bowie
GLOBAL CATASTROPHES Bill McGuire
GLOBAL ECONOMIC HISTORY Robert C. Allen
GLOBALIZATION Manfred Steger
GOD John Bowker
GOETHE Ritchie Robertson
THE GOTHIC Nick Groom
GOVERNANCE Mark Bevir
THE GREAT DEPRESSION AND THE NEW DEAL Eric Rauchway
HABERMAS James Gordon Finlayson

HAPPINESS Daniel M. Haybron
THE HARLEM RENAISSANCE
Cheryl A. Wall
THE HEBREW BIBLE AS LITERATURE
Tod Linafelt
HEGEL Peter Singer
HEIDEGGER Michael Inwood
HERMENEUTICS Jens Zimmermann
HERODOTUS Jennifer T. Roberts
HIEROGLYPHS Penelope Wilson
HINDUISM Kim Knott
HISTORY John H. Arnold
THE HISTORY OF ASTRONOMY
Michael Hoskin
THE HISTORY OF CHEMISTRY
William H. Brock
THE HISTORY OF LIFE Michael Benton
THE HISTORY OF MATHEMATICS
Jacqueline Stedall
THE HISTORY OF MEDICINE
William Bynum
THE HISTORY OF TIME
Leofranc Holford-Strevens
HIV/AIDS Alan Whiteside
HOBBES Richard Tuck
HOLLYWOOD Peter Decherney
HORMONES Martin Luck
HUMAN ANATOMY Leslie Klenerman
HUMAN EVOLUTION Bernard Wood
HUMAN RIGHTS Andrew Clapham
HUMANISM Stephen Law
HUME A. J. Ayer
HUMOUR Noël Carroll
THE ICE AGE Jamie Woodward
IDEOLOGY Michael Freeden
INDIAN CINEMA Ashish Rajadhyaksha
INDIAN PHILOSOPHY Sue Hamilton
INFECTIOUS DISEASE Marta L. Wayne
and Benjamin M. Bolker
INFORMATION Luciano Floridi
INNOVATION Mark Dodgson and
David Gann
INTELLIGENCE Ian J. Deary
INTERNATIONAL LAW Vaughan Lowe
INTERNATIONAL MIGRATION
Khalid Koser
INTERNATIONAL RELATIONS
Paul Wilkinson
INTERNATIONAL SECURITY
Christopher S. Browning
IRAN Ali M. Ansari
ISLAM Malise Ruthven
ISLAMIC HISTORY Adam Silverstein
ISOTOPES Rob Ellam
ITALIAN LITERATURE Peter Hainsworth
and David Robey
JESUS Richard Bauckham
JOURNALISM Ian Hargreaves
JUDAISM Norman Solomon
JUNG Anthony Stevens
KABBALAH Joseph Dan
KAFKA Ritchie Robertson
KANT Roger Scruton
KEYNES Robert Skidelsky
KIERKEGAARD Patrick Gardiner
KNOWLEDGE Jennifer Nagel
THE KORAN Michael Cook
LANDSCAPE ARCHITECTURE
Ian H. Thompson
LANDSCAPES AND
GEOMORPHOLOGY
Andrew Goudie and Heather Viles
LANGUAGES Stephen R. Anderson
LATE ANTIQUITY Gillian Clark
LAW Raymond Wacks
THE LAWS OF THERMODYNAMICS
Peter Atkins
LEADERSHIP Keith Grint
LEARNING Mark Haselgrove
LEIBNIZ Maria Rosa Antognazza
LIBERALISM Michael Freeden
LIGHT Ian Walmsley
LINCOLN Allen C. Guelzo
LINGUISTICS Peter Matthews
LITERARY THEORY Jonathan Culler
LOCKE John Dunn
LOGIC Graham Priest
LOVE Ronald de Sousa
MACHIAVELLI Quentin Skinner
MADNESS Andrew Scull
MAGIC Owen Davies
MAGNA CARTA Nicholas Vincent
MAGNETISM Stephen Blundell
MALTHUS Donald Winch
MANAGEMENT John Hendry
MAO Delia Davin
MARINE BIOLOGY Philip V. Mladenov
THE MARQUIS DE SADE John Phillips
MARTIN LUTHER Scott H. Hendrix
MARTYRDOM Jolyon Mitchell

MARX Peter Singer
MATERIALS Christopher Hall
MATHEMATICS Timothy Gowers
THE MEANING OF LIFE Terry Eagleton
MEASUREMENT David J. Hand
MEDICAL ETHICS Tony Hope
MEDICAL LAW Charles Foster
MEDIEVAL BRITAIN John Gillingham and Ralph A. Griffiths
MEDIEVAL LITERATURE Elaine Treharne
MEDIEVAL PHILOSOPHY John Marenbon
MEMORY Jonathan K. Foster
METAPHYSICS Stephen Mumford
THE MEXICAN REVOLUTION Alan Knight
MICHAEL FARADAY Frank A. J. L. James
MICROBIOLOGY Nicholas P. Money
MICROECONOMICS Avinash Dixit
MICROSCOPY Terence Allen
THE MIDDLE AGES Miri Rubin
MILITARY JUSTICE Eugene R. Fidell
MINERALS David Vaughan
MODERN ART David Cottington
MODERN CHINA Rana Mitter
MODERN DRAMA Kirsten E. Shepherd-Barr
MODERN FRANCE Vanessa R. Schwartz
MODERN IRELAND Senia Pašeta
MODERN ITALY Anna Cento Bull
MODERN JAPAN Christopher Goto-Jones
MODERN LATIN AMERICAN LITERATURE Roberto González Echevarría
MODERN WAR Richard English
MODERNISM Christopher Butler
MOLECULAR BIOLOGY Aysha Divan and Janice A. Royds
MOLECULES Philip Ball
THE MONGOLS Morris Rossabi
MOONS David A. Rothery
MORMONISM Richard Lyman Bushman
MOUNTAINS Martin F. Price
MUHAMMAD Jonathan A. C. Brown
MULTICULTURALISM Ali Rattansi
MUSIC Nicholas Cook
MYTH Robert A. Segal
THE NAPOLEONIC WARS Mike Rapport
NATIONALISM Steven Grosby
NELSON MANDELA Elleke Boehmer
NEOLIBERALISM Manfred Steger and Ravi Roy
NETWORKS Guido Caldarelli and Michele Catanzaro
THE NEW TESTAMENT Luke Timothy Johnson
THE NEW TESTAMENT AS LITERATURE Kyle Keefer
NEWTON Robert Iliffe
NIETZSCHE Michael Tanner
NINETEENTH-CENTURY BRITAIN Christopher Harvie and H. C. G. Matthew
THE NORMAN CONQUEST George Garnett
NORTH AMERICAN INDIANS Theda Perdue and Michael D. Green
NORTHERN IRELAND Marc Mulholland
NOTHING Frank Close
NUCLEAR PHYSICS Frank Close
NUCLEAR POWER Maxwell Irvine
NUCLEAR WEAPONS Joseph M. Siracusa
NUMBERS Peter M. Higgins
NUTRITION David A. Bender
OBJECTIVITY Stephen Gaukroger
THE OLD TESTAMENT Michael D. Coogan
THE ORCHESTRA D. Kern Holoman
ORGANIZATIONS Mary Jo Hatch
PANDEMICS Christian W. McMillen
PAGANISM Owen Davies
THE PALESTINIAN-ISRAELI CONFLICT Martin Bunton
PARTICLE PHYSICS Frank Close
PAUL E. P. Sanders
PEACE Oliver P. Richmond
PENTECOSTALISM William K. Kay
THE PERIODIC TABLE Eric R. Scerri
PHILOSOPHY Edward Craig
PHILOSOPHY IN THE ISLAMIC WORLD Peter Adamson
PHILOSOPHY OF LAW Raymond Wacks
PHILOSOPHY OF SCIENCE Samir Okasha

PHOTOGRAPHY Steve Edwards
PHYSICAL CHEMISTRY Peter Atkins
PILGRIMAGE Ian Reader
PLAGUE Paul Slack
PLANETS David A. Rothery
PLANTS Timothy Walker
PLATE TECTONICS Peter Molnar
PLATO Julia Annas
POLITICAL PHILOSOPHY David Miller
POLITICS Kenneth Minogue
POSTCOLONIALISM Robert Young
POSTMODERNISM Christopher Butler
POSTSTRUCTURALISM Catherine Belsey
PREHISTORY Chris Gosden
PRESOCRATIC PHILOSOPHY Catherine Osborne
PRIVACY Raymond Wacks
PROBABILITY John Haigh
PROGRESSIVISM Walter Nugent
PROTESTANTISM Mark A. Noll
PSYCHIATRY Tom Burns
PSYCHOANALYSIS Daniel Pick
PSYCHOLOGY Gillian Butler and Freda McManus
PSYCHOTHERAPY Tom Burns and Eva Burns-Lundgren
PUBLIC ADMINISTRATION Stella Z. Theodoulou and Ravi K. Roy
PUBLIC HEALTH Virginia Berridge
PURITANISM Francis J. Bremer
THE QUAKERS Pink Dandelion
QUANTUM THEORY John Polkinghorne
RACISM Ali Rattansi
RADIOACTIVITY Claudio Tuniz
RASTAFARI Ennis B. Edmonds
THE REAGAN REVOLUTION Gil Troy
REALITY Jan Westerhoff
THE REFORMATION Peter Marshall
RELATIVITY Russell Stannard
RELIGION IN AMERICA Timothy Beal
THE RENAISSANCE Jerry Brotton
RENAISSANCE ART Geraldine A. Johnson
REVOLUTIONS Jack A. Goldstone
RHETORIC Richard Toye
RISK Baruch Fischhoff and John Kadvany
RITUAL Barry Stephenson
RIVERS Nick Middleton
ROBOTICS Alan Winfield
ROMAN BRITAIN Peter Salway
THE ROMAN EMPIRE Christopher Kelly
THE ROMAN REPUBLIC David M. Gwynn
ROMANTICISM Michael Ferber
ROUSSEAU Robert Wokler
RUSSELL A. C. Grayling
RUSSIAN HISTORY Geoffrey Hosking
RUSSIAN LITERATURE Catriona Kelly
THE RUSSIAN REVOLUTION S. A. Smith
SAVANNAS Peter A. Furley
SCHIZOPHRENIA Chris Frith and Eve Johnstone
SCHOPENHAUER Christopher Janaway
SCIENCE AND RELIGION Thomas Dixon
SCIENCE FICTION David Seed
THE SCIENTIFIC REVOLUTION Lawrence M. Principe
SCOTLAND Rab Houston
SEXUALITY Véronique Mottier
SHAKESPEARE'S COMEDIES Bart van Es
SIKHISM Eleanor Nesbitt
THE SILK ROAD James A. Millward
SLANG Jonathon Green
SLEEP Steven W. Lockley and Russell G. Foster
SOCIAL AND CULTURAL ANTHROPOLOGY John Monaghan and Peter Just
SOCIAL PSYCHOLOGY Richard J. Crisp
SOCIAL WORK Sally Holland and Jonathan Scourfield
SOCIALISM Michael Newman
SOCIOLINGUISTICS John Edwards
SOCIOLOGY Steve Bruce
SOCRATES C. C. W. Taylor
SOUND Mike Goldsmith
THE SOVIET UNION Stephen Lovell
THE SPANISH CIVIL WAR Helen Graham
SPANISH LITERATURE Jo Labanyi
SPINOZA Roger Scruton
SPIRITUALITY Philip Sheldrake
SPORT Mike Cronin
STARS Andrew King

STATISTICS David J. Hand
STEM CELLS Jonathan Slack
STRUCTURAL ENGINEERING David Blockley
STUART BRITAIN John Morrill
SUPERCONDUCTIVITY Stephen Blundell
SYMMETRY Ian Stewart
TAXATION Stephen Smith
TEETH Peter S. Ungar
TERRORISM Charles Townshend
THEATRE Marvin Carlson
THEOLOGY David F. Ford
THOMAS AQUINAS Fergus Kerr
THOUGHT Tim Bayne
TIBETAN BUDDHISM Matthew T. Kapstein
TOCQUEVILLE Harvey C. Mansfield
TRAGEDY Adrian Poole
TRANSLATION Matthew Reynolds
THE TROJAN WAR Eric H. Cline
TRUST Katherine Hawley
THE TUDORS John Guy
TWENTIETH-CENTURY BRITAIN Kenneth O. Morgan
THE UNITED NATIONS Jussi M. Hanhimäki
THE U.S. CONGRESS Donald A. Ritchie
THE U.S. SUPREME COURT Linda Greenhouse
UTOPIANISM Lyman Tower Sargent
THE VIKINGS Julian Richards
VIRUSES Dorothy H. Crawford
WAR AND TECHNOLOGY Alex Roland
WATER John Finney
THE WELFARE STATE David Garland
WILLIAM SHAKESPEARE Stanley Wells
WITCHCRAFT Malcolm Gaskill
WITTGENSTEIN A. C. Grayling
WORK Stephen Fineman
WORLD MUSIC Philip Bohlman
THE WORLD TRADE ORGANIZATION Amrita Narlikar
WORLD WAR II Gerhard L. Weinberg
WRITING AND SCRIPT Andrew Robinson
ZIONISM Michael Stanislawski

Available soon:

AGEING Nancy A. Pachana
TELESCOPES Geoff Cottrell
CALVINISM Jon Balserak
HOME Michael Allen Fox
ROCKS Jan Zalasiewicz

For more information visit our website

www.oup.com/vsi/

David J. Hand

MEASUREMENT

A Very Short Introduction

OXFORD
UNIVERSITY PRESS

Great Clarendon Street, Oxford, OX2 6DP,
United Kingdom

Oxford University Press is a department of the University of Oxford.
It furthers the University's objective of excellence in research, scholarship,
and education by publishing worldwide. Oxford is a registered trade mark of
Oxford University Press in the UK and in certain other countries

First edition published in 2016

Impression: 7

Published in the United States of America by Oxford University Press
198 Madison Avenue, New York, NY 10016, United States of America

British Library Cataloguing in Publication Data
Data available

Library of Congress Control Number: 2016942911

ISBN 978–0–19–877956–8

Printed in Great Britain by
Ashford Colour Press Ltd., Gosport, Hampshire.

Contents

Acknowledgements

I would like to express my appreciation and thanks to Shelley Channon, Colette Bowe, Mike Crowe, and an anonymous reader for their very helpful and constructive comments on an earlier draft of this book, and to Latha Menon for recognizing the need for such a book and for guiding it on its route to publication.

List of illustrations

Chapter 1
A brief history

> 'Wherefore in all great works are Clerks so much desired? Wherefore are Auditors so well fed? What causeth Geometricians so highly to be enhaunsed? Why are Astronomers so greatly advanced? Because that by number such things they finde, which else would farre excell mans minde.'
>
> Robert Recorde (1540)

Measurement is at least as old as civilization. If we regard the beginnings of agriculture as the start of civilization, then this is true in a literal sense: agriculture requires assessing *how long* it will take and *how many* men are needed to plough a field or harvest a crop. Measurement is also the core of trade: we need to know *how much* we will get in exchange, what *volume* of beer our money will buy, how *long* a length of cloth is, what a loaf of bread *weighs*, and so on. And trade is intimately bound up with accounting. No merchant could be successful without careful control of input and output. Likewise, construction requires measurement: the Great Pyramid of Giza did not spring into existence without careful planning and measurement—how many stones and of what sizes would be needed? Moreover many of the stones fit together with extremely high accuracy, so that very precise length measurements would have been needed. And measurement is also central to navigation: how many days of travel will it take, and in what direction should we go to reach our destination?

Contrary to popular misconception, even the ancient Greeks recognized that the Earth is round. As Aristotle wrote in Book II of *On the Heavens* in 350 BC: 'since a small change of position on our part...visibly alters the circle of the horizon, so that the stars above our heads change their position considerably...This proves that the Earth is spherical and that its periphery is not large...'. Not large, perhaps, but it is generally Eratosthenes (*c.*276–194 BC) who is credited with measuring just how large it is, though he wasn't the first to try. Eratosthenes noticed that on the day of the summer solstice, when the Sun was directly overhead at Syrene (nowadays Aswan), and hence casting no shadow, it was at an angle of 7.2 degrees to the vertical in Alexandria, which he believed was about 850 km due north (I'm using modern units, of course). Relatively straightforward trigonometry (straightforward, that is, from a modern perspective) allowed him to estimate the Earth's circumference to be about 45,000 km (there is some uncertainty about the size of the *stade*, the measurement unit he used). In any case this is not bad, considering that the actual value is about 40,000 km.

The physical size of natural biological objects was frequently used as a basic unit in measurement. These often have the property that they are roughly the same size. For example, dry grains of wheat were used as elementary units of weight in the Middle Ages in England, the distance between the last two joints of the middle finger of a man's hand was used as a length measure, the fathom was defined as the height of a man, and the cubit was the length of a man's forearm, from the elbow to the tip of the extended fingers.

We can see from these examples that the Greek philosopher Protagoras's statement *man is the measure of all things* can sometimes be taken almost literally. More generally, given that measurements were developed to facilitate human interactions

1. Instructions for determining the length of the rood and the foot, by arranging for sixteen men to place their feet one behind the other.

and help everyday life, it is not surprising that units were often based on human activities. Thus, for example, we find measures based on how much a miner could dig in one day, the amount of land a team of oxen could plough in a season, and the distance that a bow could shoot an arrow.

Variability in the size of natural objects can be tackled by taking the average of several of them to yield the basic unit. This uses the statistical phenomenon that the variability of the average size of samples of objects is less than the variability between the objects themselves—because in an average the large objects tend to balance out the small ones. Here's a medieval example from a surveyors' manual written by Jacob Koebel in 1570 (see Figure 1): 'Stand at the door of a church on a Sunday and bid sixteen men to stop, tall ones and small ones, as they happen to pass out when the service is finished; then make them put their left feet one behind the other, and the length thus obtained shall be a right and lawful rood to measure and survey the land with, and the 16th part of it shall be the right and lawful foot.'

An alternative way to overcome the problem that natural variability of biological objects leads to random differences in the size of the basic unit is to adopt more fundamental physical objects. So, for example, in 1791 the French Academy of Sciences defined the metre as one ten-millionth part of the length of the distance from the North Pole to the Equator. Unfortunately, even this approach is essentially arbitrary. The Earth is not an exact sphere, and the flattening along the axis of rotation meant that the initial standard metre was a tiny bit shorter than it should have been—so, instead of the circumference of the Earth being 40 million standard metres, it is 40,007,863 m.

Once a basic unit has been chosen, larger units can be defined in terms of it: a yard is three feet, an imperial ton is 2,240 pounds, and so on. Moreover, by partitioning the basic unit, finer units can also be defined: a centimetre is one hundredth of a metre, and a second was originally defined as 1/86,400 of a solar day.

Although the notion of averaging multiple small objects to yield the basic measurement unit leads to greater accuracy, some variability remains. Furthermore, it is obvious that there is huge arbitrariness in this: if there is nothing fundamental leading to the choice of basic objects used, other systems of measurement can be adopted. Different species of crops (of wheat, for example) have different seed sizes, different races of humans have different statures, and so on. Given this, it is hardly surprising that history shows that a vast number of different measurement systems have been developed.

This can lead to difficulties. It means that one village's system for measuring the volume or weight of trade goods may be incompatible with that of the next village—with obvious problems for trade. In a word, we might say that measurement based on units which differ from place to place does not *travel.*

Writing in 1794, Arthur Young described his exasperation with this during his journeys through pre-revolutionary France five years earlier: 'the infinite perplexity of the measures exceeds all comprehension. They differ not only in every province, but in every district, and almost in every town.'

Even the same name could mean different things. In his historical survey of measurement systems, Herbert Arthur Klein remarked that 'a given unit of length recognised in Paris, for example, was about 4 per cent longer than that in Bordeaux, 2 per cent longer than that in Marseilles, and 2 per cent shorter than that in Lille', while J. H. Alexander, writing in 1850, referred to 110 separate values for the ell in Europe. And lest the reader feel tempted to self-congratulation on how far we have come since those days, she might like to recall that until 1959 (when it was internationally defined to be 0.9144 metres) the yard varied in length between countries, and that even now American acres are 0.024 square metres larger than British acres, while the US liquid pint is just over 473 ml whereas the UK pint is roughly 569 ml (which can have unexpected consequences for American visitors to the UK).

The fact that different choices were made for the object to serve as a basic unit is one reason for this diversity, but sometimes the basic unit was not any natural object of regular size, but simply a particular standard length created for that purpose. Two examples are the *toise* set in the Grand Châtelet wall in Paris, and the *bichet* held in the municipal offices of Saint-Étienne. The first of these was a standard length, made of iron, and the second a particular vessel for measuring volume of grain.

A second reason for the diversity was that the measurement unit was uniquely attached to the substance being measured, rather than to the physical property. Thus volumes of grain, wine, and coal each had different units, even though they were all measuring volume. Indeed, it is not a priori obvious that the concept of 'volume' is the same for all substances—grain can be heaped in a cup, whereas milk cannot. This also explains the existence of distinct measures of volume for grain based on whether it was heaped or 'combed', piled or flat.

Likewise, if land area is to be measured by the amount of a crop it yields, or the physical effort required to plough it, then richer land would appear to be using different units from poorer land.

From a modern perspective, it probably seems obvious that a single unified system would confer many benefits. As the French philosopher and mathematician Condorcet put it in 1793: 'The uniformity of weights and measures cannot displease anyone but those lawyers who fear a diminution in the number of trials, and those merchants who fear anything that renders the operations of commerce easy and simple.' With these sorts of merits, it was probably inevitable that a unified system would eventually be adopted. However, resistance to adopting innovations seems to be a characteristic of measurement technology. If it applies to physical measurements—to length, weight, volume, and so on—we shall see that it applies manyfold to measurement in other domains, such as in psychology and economics.

One of the earliest proposals for a unified system of physical measurements was made around 1670, when Gabriel Mouton suggested that France's many different units should be replaced by a decimal system, with increasing units being defined in multiples of ten. But acceptance was slow, and it took over 100 years before such a system was adopted for some French units, and much longer before the system was adopted more widely around the world. Indeed, outside scientific contexts, some countries—notably the USA—still resist such simplification. In the UK, on 31 December 2009, after a ten-year switching period, it became illegal for UK traders to display imperial measurements (pounds and ounces) alongside metric units. I should add that there are some exceptions, such as the pint for beer and milk, and troy ounces for precious metals. It makes you wonder if the process will ever be complete.

The ten-year process of switching to metric units in the UK had its incidents: the *Daily Telegraph* of 19 February 2002 reported that 'John Dove, who runs a fish shop in Camelford, Cornwall, was convicted of selling mackerel at £1.50 a pound. Julian Harman, also of Camelford, was convicted of selling brussels sprouts at 39p a pound.' I can only imagine the puzzled looks these miscreants must have received from other felons in the cells.

While trade was one of the main motivators for a uniform system of measurement, in recent centuries another important driver has been science. Science is perhaps the prime example of globalization, with research being publicly available in international journals, no matter what the nationality of the researchers. But even in science things can go wrong: a dramatic illustration of the consequences of using different systems of units was the 1999 loss of the Mars Climate Orbiter, when a mistaken rocket burn meant it passed too close to the planet, hitting the upper atmosphere and disintegrating. Values of force in units of pounds had been used instead of the expected newtons (N). One pound equals 4.45 N.

At the 11th General Conference for Weights and Measures in 1960 the Système International d'Units, or the SI units, was introduced. This consists of seven basic units, of length (the metre, m), mass (the kilogram, kg), time (the second, s), electric current (the ampere, A), temperature (the degree kelvin, K), quantity of substance (the mole, mol), and luminous intensity (the candela, cd). Another twenty-two named units are defined as powers and combinations of these basic seven. For example, the hertz is a measure of frequency, defined in units of s^{-1} (that is, a count per second), and the joule is a unit of energy, defined in units of

$kg.m^2.s^{-2}$. Smaller and larger units are defined as ten or one-tenth multiples of other units, and often given particular prefixes (such as kilo for 10^3=1,000 and micro for 10^{-6}=1/1,000,000). It is worth noting that the SI system has been created to be an evolving system, not one set in concrete. As measurement technology progresses, so new units and prefixes are defined.

Once a basic unit has been specified, we can make copies of it, and these can be transported around the world. Physical objects which give the standard unit size, like the toise and bichet, are called etalons, from the French *étalonner*, meaning 'to calibrate'. For the metre, a platinum-iridium bar with two scratches which were one standard metre apart when the bar was at 0 °C at standard atmospheric pressure was held at The International Bureau of Weights and Measures in Sèvres, France, and copies of it were created for countries which had signed up to the Metre Convention. A similar platinum-iridium etalon was created for the standard kilogram.

Implicit in these etalons is the notion that their length or mass remains constant. Platinum-iridium is a particularly hard and non-reactive alloy. Nonetheless, all physical objects are subject to the ravages of time, even if some change more slowly than others. For the kilogram etalon, even though it is kept under triple bell-jars to protect it from outside influences, tiny traces of microscopic materials deposit on it, and infinitesimally increase its mass over time. Likewise, length bars, even if maintained under constant conditions, shrink slightly over long periods. Of course, the magnitude of these changes is tiny, and certainly do not matter for almost all applications. But in some scientific applications one part in a million or even a billion is critical.

Since all concrete physical objects have shortcomings for measurement, an alternative is instead to define the standard units in more fundamental terms. In the first half of the 20th century, attention switched to measuring length using a count of wavelengths of light of a given frequency. Wavelengths of monochromatic light never change, and a series of such measures were proposed, including that based on the red spectral emission line from cadmium (with a wavelength of 644 billionths of a metre—that is 644 nanometres), and the reddish-orange line from krypton-86. Formally, a metre defined in this way is equal to '1,650,763.73 wavelengths in vacuum of the radiation corresponding to the transition between the levels 2 p^{10} and 5 d^5 of the krypton-86 atom'. Even this definition has its shortcomings, however, and in 1983, the metre was redefined as 'the length of the path travelled by light in vacuum during a time interval of 1/299,792,458 of a second'.

It will be apparent from this potted history that, as time progresses, so greater and greater accuracy is required of measurement procedures. This is a general principle: rough measurements are sufficient for certain purposes, but new applications often require greater accuracy. This is illustrated by the history of navigation, where, as lengthy sea voyages became more common, so more precise measurement of latitude and longitude became critical.

Latitude can be measured by relating it to angle of declination of the Sun at noon, given the time of the year. But one degree error in latitude measurement translates to an error of over 100 km in location, which means accurate measures of angle are very important.

Measuring longitude calls for a different approach. In particular it requires working out the time difference between noon at the Greenwich meridian and noon at your location (since this tells you how much the Earth rotates between these two times, and hence the angle between the two locations, and hence the distance on the Earth's surface between them). To do this, we need an accurate measurement of time; a clock which can keep regular time over long periods despite the buffeting that it will receive on the seas.

This need stimulated the offer of prizes for cracking the problem, including one from King Philip II of Spain in 1567, King Philip III of Spain in 1598, and from the UK Parliament in 1714.

A second driver of the need for improved accuracy was the industrial revolution. More sophisticated machines required more accurately produced parts, or they would grind to a halt. And this accuracy was built on careful measurement. And, lest we think it is solely relevant to mechanical contrivances, the same sort of argument applies to chemical manufacturing (precise amounts of quantities, degree of purity of constituents, etc.) and other industrial processes.

A third driver of the need for improved accuracy in physical measurements has been scientific advance. By definition, scientific advances butt against the boundaries of knowledge, and this is typically where physical phenomena are most difficult to discern: the signal may be of the same order of magnitude as the noise, so subtle and sensitive procedures are required to measure the former. A classic example is the discovery of Pluto in the late 19th century on the basis of measured perturbations of the orbit of Uranus.

More sophistication

Up until this point in our story we have only considered relatively straightforward physical measurements. While it is true that these constituted the origins of measurement, they were only the beginning.

Measuring the weight of an object is in principle straightforward: we place the object on one pan of a weighing scale and find the number of unit weight objects which just balance it. Likewise, measuring the length of an object is straightforward: we see how many copies of the unit length ruler (a foot long, say) will, when laid end to end in a straight line, reach from one end of the object to the other. Even for weight and length, however, this leaves some questions unanswered. How should we measure the weight of a skyscraper, or the distance to the Sun? Clearly more elaborate approaches will be needed.

Tribal leaders may well know everyone in their tribe, so will not need to resort to counting or elaborate censuses to get an idea of the resources they command. But rulers of larger groups—kings or emperors, for example—would need objective quantitative methods to determine how much tax they could expect to bring in, or how many men they might lead in an army.

In the 18th century there was concern about declining population numbers—though without objective evidence. Montesquieu, writing in 1721, said, 'After a calculation as exact as may be in the circumstances, I have found that there are upon the earth hardly one-tenth part of the people which there were in ancient times... if [the depopulation] continues, in ten centuries the earth will be a desert.'

The obvious solution to determining the size of a population might seem to be to count the people—to conduct a census. But this exercise is far from straightforward. First, it requires people to stay put while the counting is going on, so that they do not get overlooked or counted more than once. And, secondly, and more problematically, populations change. People migrate between countries, die, and are born. Modern censuses are very elaborate exercises costing hundreds of millions of dollars.

For these reasons, alternative ways of estimating population sizes had to be developed. A popular strategy was to use a *multiplication factor*. So, for example, a count of the fireplace hearths in an area was a common proxy for the local number of families—and hence the local tax to be collected. If each hearth serviced, say, five people, the total population size could be estimated by multiplying the number of hearths by five. Likewise, taking the number of births in a year from parish records and multiplying by, say, 25, could provide an estimate for the population in that neighbourhood. It is probably obvious to the reader that the accuracy of these methods is questionable: on the grounds of the choice of multiplier, certainly, but also on the grounds that some people may well try to avoid being included in records related to tax, or because the birth records were unreliable. These are not merely ancient problems: a 2013 UNICEF report noted that the births of nearly 230 million children under the age of 5 had never been registered.

The case of population size illustrates something deeper. The number of inhabitants of a town or country is not a property of the individuals themselves, but of the town or country. It is a characteristic of the higher level entity constituted from an aggregation of lower level entities. Looked at from this perspective, it becomes clear that there are all sorts of other characteristics of aggregate phenomena which emerge only at higher levels. So, for example, we have the *crime rate* within a city and the *unemployment rate* of a country. Indeed, it has been argued that these developments in notions of measurement are one of the drivers behind the creation of the concept of *society*. As Ken Alder has elegantly put it: 'Measures are more than a creation of society, they *create* society.' This is an important notion, because it provides us with our first glimpse of the remarkable power of measurement: new concepts are invented, which require measurement, and these concepts are attributes of newly defined objects.

The step from measuring concrete attributes like length and weight to measuring characteristics such as crime rate is perhaps the first step on a ladder of increased abstraction in measurement. Still within the social and economic domain, we can contemplate measuring economic progress, for example through Gross Domestic Product (GDP) and Gross National Income (GNI), as well as inflation rate. It is immediately clear that measuring these sorts of attributes requires procedures very different from measuring length or weight. At the very least, we cannot conceive of something which has a basic unit of inflation, multiple copies of which can be placed end to end (as for length) or together on a balance (as for weight), to measure the rate of inflation of, say, the UK. Something different—and more advanced—is needed.

Contemplation of what is needed shows that implicit in many measurement procedures is the very definition of what is being measured. In such cases, the two notions—what the attribute *is* and how it is *measured*—are two sides of the same coin. This is clearly illustrated with price indices such as measures of inflation. There are many different price indices. For example, in the US we have the CPI (Consumer Price Index) and variants such as the CPI-U and the CPI-W, and in the UK we have the CPI, CPIH, RPI (Retail Price Index), and RPIJ. Different measures differ in various ways, including the population they are intended to apply to, the goods and services which contribute to the overall measure, and the way the different components are statistically combined. These differences are not surprising—different measures have been developed for different purposes—some as cost of living indices, others as macroeconomic indicators, and so on.

In length and weight measurement we assigned numbers to different objects so that the relationship between the numbers corresponded to the relationship between objects. For example, if one object was heavier than another, as indicated by the fact that it tipped a set of weighing scales down, we assigned a larger *weight* number to that object. If two objects exactly balanced a third, we assigned weight numbers to the three objects so that the sum of the two equalled the third. This sort of approach to measurement is called *representational* measurement: we are representing the objects and their relationships by numbers and their relationships.

In contrast, for the social science and economic examples, we construct a measure (e.g. from the prices of the goods people purchase) which has the right sort of properties for our intended use. This sort of measurement is called *pragmatic* measurement.

Pragmatic measurements also occur in many other domains, not merely in socio-economic areas. In psychology, for example, we are often called upon to measure subjective phenomena. Take pain as an example. The most obvious way for me to determine the extent of pain you feel is by asking you. However, as we shall see, simply asking people can lead to inaccurate results. Precisely because of this, a huge amount of research effort has been put into developing accurate and reliable measures of phenomena such as pain, depression, wellbeing, quality of life, and so on, with the result that some very good measures are now available. Clearly such measures are heavily pragmatic: for example, the definition of what is meant by wellbeing and the way in which it is measured are intimately intertwined.

Many measurement procedures make use of a combination of representational and pragmatic concepts, so it is best to think of these perspectives as being extremes of a continuum. Medicine, in particular, makes extensive use of both ideas. On the one hand we have measures such as Forced Expiratory Volume (FEV), an explicit measure of how much air someone can exhale in a forced breath. This is clearly a very representational measurement—it is a direct mapping from a physical phenomenon to numbers. And towards the other extreme we have something like the Apgar score, used to assess the state of a newborn baby. This measure is defined as the sum of five factors—skin colour, heart rate, reflexes, muscle tone, and breathing rate—each scored 0, 1, or 2. Clearly the definition of what is measured by the Apgar scale and how it is measured are really one and the same—it is a pragmatic measurement, designed for a particular purpose, rather than a representational measurement mapping an aspect of the world to a numerical representation.

We have seen how measurement spans the entire range of human interests, from simple physical phenomena, through social constructs, to understanding the human brain and what people think and experience. Perhaps not surprisingly, and in common with all advances, progress in measurement technology has not been without resistance. This has been the case for physical measurements (e.g. Robert Harrington, writing in 1804: 'Such pretensions to *nicety* in experiments of this nature, are truly laughable! They will be telling us some day of the WEIGHT of the MOON, even to *drams*, *scruples*, and *grains*...'). And also the case for medical and psychological measurements (e.g. Richard Shryock: 'Measurement, declared so distinguished an authority as Goethe, could be employed in strictly physical science, but biologic, psychologic, and social phenomena necessarily eluded the profane hands of those who would reduce them to quantitative abstractions'). Unfortunately, despite the awesome advances in all areas of human endeavour, such resistance is still encountered in some quarters—often in the mistaken belief that the extra information that measurement brings somehow reduces our understanding and depth of experience.

The first half of the 20th century witnessed an intensive debate about the nature of measurement. This period also saw intensive debate about the relationship between statistical methods and measurement: if a measurement does not correspond to a natural way of combining objects (as, for example, there is no equivalent to putting two 'intelligences' together, like putting two objects together on the pan of a weighing scales), then what sense does it make to add the numerical values together when we calculate an average intelligence? We look at this question in more detail in Chapter 7.

Although I have given examples of people finding it difficult to come to terms with progress in measurement technology, there are also plenty of examples of the contrary. In 1891, William Thomson (later, Lord Kelvin) famously said: 'I often say that when you can measure what you are speaking about and express it in numbers you know something about it; but when you cannot measure it, when you cannot express it in numbers, your knowledge of it is of a meagre and unsatisfactory kind...'. More recently, psychologist Donald Laming wrote: 'If the sums do not add up, the science is wrong. If there are no sums to be added up, no one can tell whether the science is right or wrong.'

One thing is clear. Notions of measurement are pervasive—even in our very language. We speak of the measure of a man, his strength of character, the greatest good for the greatest number. We measure how long we live, how tall we are, how effective is a diet, how far we have to travel, and how much it all costs. Our modern world-view is constructed on a framework of measurement. Measurements both reflect structure in the natural world, and impose structure upon it. You might say that we see the world through the spectacles of measurement.

Chapter 2
What is measurement?

In Chapter 1 we saw that measurement procedures could be placed on a continuum which stretched from *representational* at one end to *pragmatic* at the other. Extreme representational measurement involved establishing a mapping from objects and their relationships to numbers and *their* relationships. Pragmatic measurement involved devising a measurement procedure which captured the essence of the characteristic of interest, so that pragmatic measurement simultaneously defined and measured the characteristic. We could almost say that representational measurement is based on *modelling observed* empirical relationships, while pragmatic measurement is based on *constructing* attributes of interest. Most measurement procedures have both representational and pragmatic aspects.

Note that even measurements of physical attributes involve a pragmatic aspect: such measurements and their application in science, engineering, and life in general, require a choice of unit. There is nothing in nature which allows us to choose between units, so the choice must be based on pragmatic considerations. To take an exaggerated example, if I wanted to study the impact of childhood diet on adult height, pragmatic considerations would lead me away from measuring height in light years, where a difference of 0.0000000000000000027 light years would matter (this is one inch, expressed in light years).

In other chapters we will see many examples of measurement procedures lying at different points on this continuum. But first, so that we can appreciate those examples more fully, we shall dig more deeply into the representational and pragmatic aspects of measurement.

Representational measurement

To illustrate how representational measurement assigns numbers to objects so that the relationships between the numbers are the same as the relationships between the objects, I shall use the simple physical example of measuring length. In particular, to strip away irrelevant and potentially confusing details, I shall assume we have a collection of sticks (these being the objects) and we want to find numbers which represent their lengths.

The first thing we notice is that, if we place one end of each stick against a wall, so that they are projecting out perpendicularly from the wall, some sticks will reach further than others. We say that some sticks are longer than others, and we can assign numbers to the sticks so that sticks which stretch further are given larger numbers. We have *represented* the order relationship between the length of the sticks by the order relationship between the numbers we have assigned.

But we can go further than this. We might notice that, if we put two of the shorter sticks end-to-end in a straight line sticking out from the wall, together they reach the same distance from the wall as one of the longer sticks. The *end-to-end concatenation* of the two shorter sticks has the same length as the longer stick. With some care, we can then choose numbers so that the sum of the numbers we assign to the two shorter sticks is equal to the number we assign to the longer stick, for any group of three sticks in our collection.

In fact, we can take the shortest stick in our collection, and determine how many copies of it, concatenated end-to-end in a

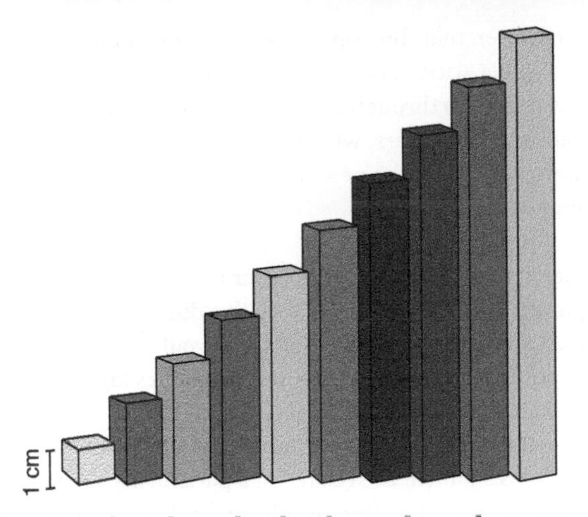

2. Cuisenaire rods can be used to show how end-to-end concatenation is mapped to addition.

straight line, stretch the same distance from the wall as any other stick we care to take (at least, approximately). We can then assign, to any stick, a number equal to the number of shorter sticks which stretch the same distance—and call this number the measured length. Figure 2 shows a set of Cuisenaire rods, which can be used in this way to introduce the notions of length measurement.

By this means we have constructed a numerical representation of our system of sticks which has the properties that (a) longer sticks have larger numbers and (b) the numbers assigned to a concatenated set of sticks add up to equal the number assigned to a longer stick which stretches the same distance. If this sounds rather theoretical, we can give a name to the length of the shortest stick, and call it our basic unit of length measurement (an inch, say) and the reader may recognize this as the standard procedure we go through when we measure the length of something. We simply see how many copies of our basic unit, our inch, stretch as far as the thing we want to measure.

Note, however, that this representation is not unique. If we found another stick, shorter than the shortest one we previously used, we could go through the same operation and assign a different set of numbers, which also had properties (a) and (b). Again we might give the new shortest stick a name—a centimetre, say.

In general, there is an infinite number of systems of numerical assignments we could make, each with a different basic unit (shortest stick). But all of these systems would be equally legitimate, in the sense that they had properties (a) and (b).

The reader may have spotted that the different systems of numerical assignment are related in a particularly simple way: we can get from one to the other just by multiplication of all the numbers in one system by a fixed value. Suppose, for example, that we have assigned numbers using the inch as our basic unit of length. To get to the numbers we would have assigned if we had used the centimetre as our basic unit, all we have to do is multiply all the numbers by 2.54, which is the number of 1-centimetre sticks which stretch 1 inch. So, if one of our sticks is 10 inches long, all we have to do is multiply by 2.54, and we have 25.4 as the length of this stick in centimetres.

Multiplication by a positive constant is a *rescaling* transformation. If we multiply the original measurements (in inches) by 1/12, for example, we are rescaling to produce measurements in feet, and so on. For length measurement, rescaling transformations are called *permissible* (sometimes *admissible*) because they lead us from one legitimate numerical representation to another. Legitimate here means that the representation has properties (a) and (b). In general, permissible transformations are just those transformations of the numbers which lead to new sets of numbers which have the same relationships as the originals, so that they also accurately represent the empirical relationships.

We can also see from this that not all transformations will preserve the relationships—so not all transformations are permissible. For example, replacing the numbers by their squares won't work. Suppose we have three sticks, which we have measured to be 1 unit, 3 units, and 4 units in length (again, think of the units as inches if you like). If we transform these numbers by squaring them, we find that order is preserved ($1 < 9 < 16$) but the addition relationship is not ($1 + 9 \neq 16$). Squaring is not a permissible transformation for length measurement.

Measurement systems for which rescalings are the permissible transformations, like length, are very important. They are ubiquitous in the physical sciences (think of mass, weight, time intervals, speed, and so on), and were the earliest kind of measurement systems adopted. The measurement scales which result from such systems are called *ratio* scales: the ratio between two lengths is the same whatever units it is expressed in. The ratio between the lengths of a 3-inch stick and a 4-inch stick, $3/4 = 0.75$, is the same as the ratio we get if we convert each length to centimetres (by multiplying each length by 2.54): $7.62/10.16 = 0.75$.

But not all measurement systems are like this. A simple example will show that there might be more to it. The *Daily Telegraph* of 8 February 1989 commented that 'Temperatures in London were still three times the February average at 55 °F (13 °C) yesterday.' (Here, °F means degrees Fahrenheit and °C degrees Celsius or Centigrade.) Given this information, you might ask what does that make the February average? Obviously it's a third of the given temperatures: it's $55/3 = 18\frac{1}{3}$ °F or $13/3 = 4\frac{1}{3}$ °C. The problem is that $18\frac{1}{3}$ °F is below freezing, while $4\frac{1}{3}$ °C is above freezing. Something seems to have gone wrong.

What has gone wrong is that rescaling transformations between the Fahrenheit and Celsius temperature scales doesn't preserve the empirical relationships: rescaling transformations are not permissible transformations for these temperature scales. Rather more elaborate transformations are needed to get from one of these temperature scales to another. There *is* a rescaling involved: the Fahrenheit scale has 180 degrees between the freezing point and boiling point of water, while the Celsius scale has only 100, so a degree on the Celsius scale is 1.8 times as large as a degree Fahrenheit. But there is also a shift of the zero point (the freezing point of water is 32 °F, but 0 °C). This means that, to get from a temperature of t °F to the corresponding temperature in °C we have to adjust for both the different sizes of the degrees and the different baselines. First we subtract the temperature of the freezing point of water to make the baselines equal (to 0); then we multiply by an appropriate factor to take account of the different size of degrees in the two scales; then we add the freezing point of water back in the new scale. So the transformation from °F to °C has the following steps:

Subtract the freezing point of water in °F: $(t - 32)$

Rescale to allow for different degree sizes: $(t - 32) \times 100/180$

Add the freezing point in °C: $(t - 32) \times 100/180 + 0$ (since 0 is the freezing point in °C).

Overall, then, to convert a temperature t in °F to one in °C we transform t using

$$(t - 32) \times 100 / 180$$

This simplifies (approximately) to the equation $0.556 \times t - 17.778$.

This equation is an example of a *linear* transformation. As well as rescaling (multiplying by 0.556) it also adds a constant (–17.778). Measurement scales in which linear transformations are the permissible transformations are called *interval* scales. For interval scales, adding *differences between* the assigned numbers preserves empirical relationships: if object 1 is x °F hotter than object 2, and object 2 is y °F hotter than object 3, then object 1 is $x + y$ °F hotter than object 3, and this addition relationship also holds in °C.

This has solved the problem arising in our original example: when dividing by 3 to get the February average, we must avoid dividing the baseline by 3. So, for °F, the February average is not 55/3 °F, but (55 – 32)/3 + 32 = 39.7 °F, while for °C it is (13 – 0)/3 + 0 = 4.3 °C, and these are equal since 0.556 × 39.7 – 17.778 = 4.3 (which, incidentally, is above freezing).

We can also see from this example that ratio scales are the special form of interval scale in which the constant to be added is 0, regardless of the units. Put another way, ratio scales are appropriate for those systems which have a natural zero point. Length and weight, for example, have a zero value, below which it is not possible to go.

The point of this temperature example is to show that not all measurements have the form of ratio scales. It means that the scope for what measurement means, and for what can be measured, is broadened.

One of the earliest researchers to recognize this was the psychologist S. S. Stevens, who identified four types of measurement scale:

- *nominal scales*, in which the only empirical property reflected by the numbers is that objects have different values of the attribute. Hair colour is an example of such an attribute: blond, black, brown, and red do not have a natural order, so one cannot say that one colour is 'greater than' another, and one cannot concatenate hair colours. Some people regard this as too limited to be dignified with the term 'measurement'.
- *ordinal scales*, in which the only relationship between objects (in terms of the empirical attribute in question) represented by the

relationship between numbers is their order relationship. For example, the Mohs scale of hardness ranks ten materials (ranging from talc, the softest, to diamond, the hardest) according to whether they can scratch or can be scratched by other materials. From a representational perspective, only the order property represents a property of the physical elements, and any set of ten numbers which had the same order could be used. It makes no sense to add two numbers together since there is no meaningful notion of 'concatenating' objects to yield a harder object. New materials can have their hardness measured by seeing where they lie on this scale. In fact, the Mohs scale uses the (pragmatic choice of the) integers 1 to 10 from softest to hardest. Another example is the 'sea state' scale, with the World Meteorological Organization defining a thirteen-point scale ranging from 0 (flat calm) to 12 (hurricane).

- *interval scales*, which we just met in the form of temperature scales, in which both the order relationship and the concatenation of differences between objects is reflected in the relationships between numbers.
- *ratio scales*, which we have also already encountered, have both order relationship and concatenation of objects reflected by the relationships between numbers.

Sometimes other scale types are also defined. For example, straightforward counting is sometimes regarded as an *absolute* scale, since it has no permissible transformations to alternative sets of numbers. *Difference* scales have permissible transformations $x \to x + a$, with a constant. In some areas of psychology *log-interval* scales are used, in which the permissible transformations have the form $x \to ax^b$ with a and b positive constants.

Some deep mathematics has been developed, exploring the relationships between the properties of empirical systems and the nature of the numerical systems which may be used to represent them, along with the permissible transformations of the numerical systems. Considerable advances have been made since the first half of the 20th century, when Stevens was doing his work, but it is interesting to note that his conclusions still essentially hold.

Simple concatenation operations, such as placing two sticks end-to-end, or two weights on the same pan of a weighing balance, are naturally mapped to addition. But more elaborate empirical relationships can also be mapped to addition. One important example leads to *conjoint measurement*. This is widely used in commercial product development, marketing, and some areas of psychology.

Conjoint measurement applies to situations in which objects each have several attributes, and where the objects can be ordered. For example, we might be able to arrange a collection of foodstuffs in order of preference, and the foodstuffs might be characterized in terms of their crunchiness, sweetness, texture, and strength of taste. Then, under certain circumstances it is possible to derive numerical scales for each of the distinct attributes—mapping different levels of crunchiness to different numbers and so on—such that the sum of the values of the attributes correctly orders the objects.

In our stick example, we chose to represent an empirical concatenation operation by addition. We put two sticks together, end-to-end, and assigned numbers to the two sticks so that the sum of the numbers was equal to the number we assigned to the stick which had the same physical length as the end-to-end concatenation. For weight, we might place two smaller objects on one pan of a weighing scale, and balance them with a single large object on the other, assigning numerical weights so that the sum of the two numbers assigned to the objects on one side equalled the number assigned to the third. In each case, we represent the empirical relationship by addition. But it's not mandatory to use addition.

We could, for example, use multiplication. So, instead of assigning numbers to our sticks so that, when we put two together, end-to-end, the numbers *added* to equal the number assigned to the single stick that stretched the same distance, we could choose our numbers so that they *multiplied* to give whatever number we assigned to the longest stick.

Again suppose we have three sticks, of three different lengths, and such that the end-to-end concatenation of two of them stretches the same distance as the third. Under the addition approach, we might assign the numbers 1, 3, 4 to them in order of length. Note that $1 < 3 < 4$ and $1 + 3 = 4$, preserving both the order relationship and the end-to-end concatenation relationship. However, we could alternatively assign the numbers 2, 8, 16, consecutively, to the three sticks. The order property is again maintained, $2 < 8 < 16$, but now, in place of addition, we have $2 \times 8 = 16$: the product of the lengths of the two shorter sticks equals the length of the longest one.

This might seem a bit odd, but that's simply because it is so common to use the addition representation. There is nothing in nature which says we must do that, and in fact the use of multiplication instead of addition to represent concatenation of objects is quite common in some domains (e.g. certain classes of statistical models). The choice of the addition representation is, in fact, a pragmatic choice: it's convenient for certain purposes.

Now, since both the additive assignment of numbers and the multiplicative assignment represent the same underlying physical properties (the order and concatenation of sticks) they must be equivalent in some sense. (If A is equivalent to B, and C is also equivalent to B, then A is equivalent to C.) And, indeed, it is easy to see that there is a mapping of the numbers which takes those used in the additive representation to those used in the multiplicative representation: simply raise 2 to the power given by the additive numbers (that is, $2^1 = 2$, $2^3 = 8$, and $2^4 = 16$). To map things the other way, from multiplicative to additive, take logs to base 2 (that is, $\log_2(2) = 1$, $\log_2(8) = 3$, $\log_2(16) = 4$).

In various contexts, numerical operations other than addition (or multiplication) are quite common. Take relativity, for example. In Newtonian physics, we combine velocities using addition: if I am walking at x km per hour on a train which is moving at y km per hour, my velocity relative to the ground is $z = x + y$ km per hour. But in relativity velocity measurements are combined using a more complex arithmetic operation:

$$z = (x + y) / (1 + xy / c^2), \text{ where } c \text{ is the speed of light.}$$

In summary, we can represent the physical magnitudes of attributes of objects by numbers, where the relationships between the numbers represent the relationships between the objects. We can use addition to represent the physical relationship, but we can also use other numerical relationships.

At this point the reader might be wondering if all this theory is necessary. The ancient Egyptians built the pyramids without worrying about permissible transformations or legitimate numerical representations. The fact is, however, that the deeper understanding provided by this theory equips us with some very powerful tools for teasing apart the secrets of the universe.

Pragmatic measurement

Since pragmatic measurement simultaneously defines the attribute being measured and specifies how to measure it, pragmatic measurement is closely related to the philosophical position of operationalism, which the Nobel laureate physicist Percy Bridgman characterized by saying 'the concept is synonymous with the corresponding set of operations'. Let us take the measurement of individual wellbeing as an example. One strategy is to formulate a single question which taps into what people mean by wellbeing. Thus we might ask, 'Thinking about your life as a whole, how satisfied are you with it?' (from America's *Changing Life* survey), or 'Would you describe yourself as very happy, somewhat happy, . . . ?' (from the Canadian *General Social Survey*), or 'On the whole, are you very satisfied, fairly satisfied, . . . , with the life you lead?' (from the *Eurobarometer*).

The different questions overlap in sense—which is reassuring, since it means they are tapping into related concepts. On the other hand, the differences between them mean they are tapping into slightly different aspects of wellbeing. That is, they are measuring slightly different things. Precisely what they are measuring is defined solely by the questions themselves. In contrast to the representational measurement case there is no more fundamental 'reality' which can be appealed to.

A rather more elaborate strategy for pragmatic measurement is to combine multiple items to produce a single score—the Apgar scale provided an example. Combining multiple aspects has several advantages over asking a single question. It means we can be confident that all aspects of the attribute are covered by the measurement procedure. It also means we have control over how the different aspects are combined and the relative weight or importance given to each. Moreover, as mentioned in Chapter 1, for statistical reasons explicitly combining multiple components to yield a single overall measure is likely to lead to a more accurate and less erratically variable result than simply asking a single question. And, finally, there is evidence, at least in certain contexts (such as measuring wellbeing), that component questions are less susceptible to systematic bias than overall questions.

Broadly speaking, there are two strategies for combining multiple items to yield a single overall measure. They have sometimes been called the *clinimetric* approach and the *psychometric* approach, because of the domains in which they are heavily used. The clinimetric approach is the more straightforward and is more overtly pragmatic: we simply choose relevant components and decide how to combine them. Again the Apgar score provides an example, with values of 0, 1, or 2 for each of skin colour, heart rate, reflexes, muscle tone, and breathing being combined (in fact, added) to yield a single score of a baby's health state. In her original paper describing this score, Virginia Apgar, the inventor of the Apgar scale, said: 'A list was made of all the objective signs which pertained in any way to the condition of the infant at birth. Of these, five signs which could be determined easily and without interfering with the care of the infant were considered useful. A rating of zero, one or two was given to each sign depending on whether it was absent or present. A score of ten indicated a baby in the best possible condition.' This is clearly pragmatic. Even if the components may involve representational measurement (heart rate lies on a ratio scale, for example), there is nothing in nature determining which items to include and how to combine them, nothing we are explicitly representing when we make those choices. Indeed, we could have chosen to include different component attributes and to combine them in ways other than addition. The measurement scales resulting from such alternative choices would be equally legitimate—albeit different. Quite clearly in this example what we are measuring is defined by how we construct the measurement procedure.

In the alternative, psychometric, approach to combining multiple components, the components are each assumed to have some explicit relationship with the concept to be measured. Another way of saying that is that we construct a theory, or model, which we believe relates the thing we want to measure (the *latent variable*) to the things we can measure (the *manifest variables*). Informally, such an approach is sometimes called *indirect* measurement. The fact that there is some sort of theory underlying this approach shows that it is in part representational—it is modelling some postulated underlying reality—but note that the choice of which tests to include is a pragmatic one, based on what the researcher thinks are relevant aspects. A classic example of the psychometric approach to combining items arises in the measurement of general intelligence, based on measurement of such things as scores on arithmetic tests, verbal tests, visuo-spatial reasoning tests, and so on. We explore this in more detail in Chapter 5.

With my co-author Peter Fayers, I have previously described psychometric approaches as 'attempting to measure a single attribute by using multiple items' and clinimetric approaches as 'attempt[ing] to summarise multiple attributes with a single index'. We went on to say that the latter 'of course, cannot be done'. But that is the essence of the pragmatic combination of attributes: finding a *useful* way to summarize multiple distinct characteristics, as the Apgar scale illustrates.

Merely because there is considerable freedom in defining a pragmatic measurement procedure does not mean 'anything goes'. To be useful, the measurements have to have good properties, such as precision and degree of replicability. The quality of the procedure may also depend on how well the resulting numbers cohere with a body of theory relating them to the results of other measurement procedures, but this is not essential. It might be that there is deep theory, involving elaborate constructs relating the results of different measurements—for example, a psychological theory relating intelligence to social wellbeing and achievement. But, on the other hand, a pragmatic measurement might 'merely' be useful because it yields accurate prediction. For example, we could build an empirical model to predict who is likely to default on mortgage repayment, purely on the basis of correlations observed in the past between various characteristics of borrowers and whether they defaulted. This would be devoid of psychological theory but it would be (and such models indeed are) extremely useful—because they can be highly accurate. This type of pragmatic measurement, constructed by finding functions of easily measurable characteristics which are highly correlated with some outcome, is particularly widespread in business and management applications.

It will be clear from this that there is a great deal of flexibility in the way pragmatic measurement procedures can be defined. The key point is that such measurement methods are chosen (and hence defined) for some purpose, and that these purposes may be various and different. This is illustrated by the existence of a large number of pain scales, each tapping into pain in a slightly different way, and measuring slightly different aspects of it. Or, as we might say, crystallizing exactly what is meant by 'pain' in slightly different ways. The multiple aspects of pain include intensity, duration, distribution, and location, and the approaches to measurement include self-report, physiological response (such as respiration rate, heart rate, blood pressure, and perspiration), brain scans, and behavioural aspects. All of these differences can have consequences in terms of diagnosis, prognosis, and treatment, so it is important to use the right measurement instrument.

Another example of pragmatic measurement, from a completely different domain, is price inflation—the rate at which prices increase. Difficulties in measuring inflation arise from the fact that goods have different prices in different outlets, the prices change differently, people buy different goods in different quantities, the nature of goods changes (a laptop bought five years ago will be very different from one bought now, even if they cost the same), and so on. Furthermore, price inflation is an *aggregate* phenomenon, relating to overall prices in a country, so somehow the different components have to be averaged—and there are different ways to do this (e.g. arithmetic vs geometric means). All of these issues, and others, can be addressed in different ways, so there are various distinct measures of inflation—all yielding different answers. It is worth stressing that this is not because some are 'more right' than others, but simply that the different measures have subtly different meanings for what 'inflation' really is.

Other familiar economic pragmatic measurements are stock market indices, seeking to summarize the overall movements of a market—such indices as the FTSE100, the Dow Jones Industrial

Average, and the S&P 500. Typically such indices are calculated as weighted averages of the prices of the individual stocks comprising the market. Pragmatic choices include which stocks to include in the index, and how to weight them.

The reader would be correct if she inferred from these examples that pragmatic measurement procedures are more common in the human sciences than the physical sciences. This is partly because of the inherent complexity and diversity of the material studied in the former. But pragmatic measurement does also arise in the physical sciences. For example, I described the Mohs scale as a way to measure hardness. This is based on minerals' relative ability to scratch each other. But other approaches are based on degree of indentation when subjected to a blow (e.g. the Brinell, Rockwell, Vickers, and Shore scales), and on the extent to which a diamond-tipped hammer will rebound from a material. All of these are measuring 'hardness', but each measurement procedure implies a slightly different meaning for hardness.

The notion of permissible transformations arose in representational measurement because of the explicit mapping from an empirical structure in the real world to a numerical representation. If different scales represent the same real-world relationships, they must also be related to each other. The permissible transformations show how they are related. In contrast, since pragmatic measurements do not have such an explicit mapping from empirical relationships in the real world, there are no constraints on the transformations that can be applied. On the other hand, what this also means is that a transformed version of a scale does not represent the same attribute as the untransformed version. Put another way, arbitrary transformations of a pragmatic scale are allowed, but each such transformation leads to a different definition of whatever is being measured.

It is often convenient, in pragmatic measurement, to arrange the resulting scores so that they have particular properties. This can aid in interpretation. Thus, for example, we might produce scores which lie between 0 and 1, or perhaps produce scores which have a particular statistical distribution for some defined population. IQ scores are an illustration of the latter, arranged to have a Gaussian or normal distribution, with a mean of 100 and standard deviation of 15. Any transformations involved in producing scales with particular properties should be regarded as part of the definition of the pragmatic measurement procedure. So, for example, the Apgar scale yields scores from 0 to 10, but we could transform it—for example, by scaling the numbers, yielding scores from 0 to 100. If we did that, we should really give it a different name and regard it as a different measure.

As we might expect, pragmatic measurement has stimulated considerable debate and controversy around the extent to which we can regard the measured attributes as 'real'—so-called *reification* of the attributes. Does defining a measurement procedure which produces consistent and replicable results, which are related in a useful way to other measures, mean that the attribute being measured is 'real'? The biologist Stephen Jay Gould criticized interpreting the *g* factor extracted in intelligence research as real:

> The idea that we have detected something 'underlying' the externalities of a large set of correlation coefficients, something perhaps more real than the superficial measurements themselves, can be intoxicating. It is Plato's essence, the abstract, external reality underlying superficial appearances. But it is a temptation that we must resist, for it reflects an ancient prejudice, not a truth of nature.

In a related vein, the philosopher John Stuart Mill said, in a footnote to an edited book by his father, 'the tendency has always been strong to believe that whatever received a name must be an entity or being, having an independent existence of its own. And if no real entity answering to the name could be found, men did not for that reason suppose that none existed, but imagined that it was something peculiarly abstruse and mysterious.'

While the debate about reification is particularly relevant in the human sciences, it is not unique to them. Gravity and magnetism cannot be observed directly, but only via their effect on other bodies, so are they themselves real?

One answer to this question might be found in whether or not different ways of measuring something lead to the same result—like measuring the weight of an object by using a weighing balance or by seeing how far a spring supporting it is stretched. If different procedures do lead to the same result, if the measurement operations are *convergent*, then we might regard the attribute as real, and having an external existence beyond the measurement procedure. Clearly representational procedures satisfy this: by definition there must be something to be represented. On the other hand, if there is only one way to measure an attribute, then it might better be seen as a construct—and pragmatic measurements would fall into that category.

Scaling

Although we have drawn a clear distinction between representational and pragmatic measurement, in many, perhaps most, cases measurement consists of a mixture of these two extremes. To take an example, consider a scale in which a respondent has to give one of five possible answers: strongly disagree, disagree, undecided, agree, strongly agree. The order of these response categories is clearly intended to represent something empirical, so the scale is representational in this respect. On the other hand, a particular choice for the category scores—say 1 to 5 respectively—would be pragmatic. There is, for example, nothing empirical being reflected in the fact that the second category has twice the score of the first. Nevertheless, if researchers consistently use the scores 1 to 5, the resulting scale can be valuable.

Single questions can be susceptible to a high degree of random variability, and, as we have already observed, scale properties can be improved by asking multiple questions and combining them. So, for example, the widely used *Likert scale* approach asks multiple related questions (or *items*), with response categories which may be scored using the integers (1 to 5, for example). These scores are then summed or averaged to yield an overall score. Note that, implicit in this procedure is the assumption that each item is of equal difficulty or importance. More elaborate measurement procedures weight them differently before adding or averaging. For obvious reasons, the Likert scale is an example of what is called a *summated rating scale*.

A criticism of using integers for the response levels in such scales is the implicit equal intervals between the levels, and many people prefer to use scores other than integers for the levels. In *Thurstone equal appearing intervals scaling*, for example, each of a number of items is given a numerical score derived by summarizing the individual scores of a number of judges. The judges initially score each item according to how much they agree with them, on a scale ranging from 1 to 11. Each item is then given an overall score equal to the median of the scores given by the judges. A subset of items is then selected which have equally spaced medians. To calculate a respondent's score using this scale, the respondent is asked which items they agree with, and the mean item score for these is taken to be the respondent's score.

A rather different strategy for choosing scores for the levels of a single question is *optimal scaling*. Here numerical values are found which optimize a relationship with another scale. So, for example, in relating income to position on a left/right political continuum, we might choose the levels of position so as to maximize the correlation between the two characteristics in some population, subject to the constraint of preserving order. These sorts of approach have been developed in great generality and sophistication. They can also be used with nominal scales—scales in which the response categories do not even have an order. For example, we might conjecture that (the intrinsically unordered characteristic) hair colour was related to a particular medical condition. Then the ratio of the proportion of people with the condition who have blonde hair to the proportion of people without the condition who have blonde hair could be used as a numerical score. In fact, the logarithm of this ratio—called a *weight of evidence*—is widely used in diagnostic systems.

Optimal scaling chooses a criterion relating the scale to one or more other variables, which is then optimized by choosing scale values for the response categories. In doing so, the properties of the scale are a consequence of the relationships—we are looking outside the continuum of the particular measurement in question. An alternative strategy is to suppose that the discrete ordered scale values (which might originally be represented by the integers 1 to 5, for example) have in fact arisen by segmenting a continuous underlying interval or ratio measurement scale. Suppose, for example, we imagined that the scores on the notional underlying measurement had a standard Gaussian distribution. Then, to illustrate, if 30 per cent of the respondents had a score of 1, we might replace that score by −0.52, since 30 per cent of values from a standard Gaussian distribution are less than −0.52. Since, in many situations, empirical phenomena do have roughly Gaussian distributions, this may be a reasonable strategy. It is, however, not without its dangers. In particular, the transformation implicit in moving from the scores 1, 2, 3, 4, and 5 to the quantiles of the Gaussian distribution can lead to inversion of the order of relationships: it is possible that the average score of one population of participants is greater than the average score of another population when using the original integer values, but is less when converted to Gaussian scores.

The items in a Likert scale are all treated as equivalent, and sum to yield the overall score. An alternative strategy is to ask a series of questions of increasing difficulty. In principle, a respondent should be able to answer all questions below a certain level of difficulty correctly, while not being able to answer the remainder. The transition point between correct and incorrect can serve as a score. This sort of strategy is used in *Guttman scaling*, where elaborate methods have been developed for ordering questions and respondents.

Guttman scaling supposes a rank order to the items or questions. *Coombs scaling*, used in political science and other areas, is based on rankings of items or statements produced by each participant. The method yields a score for each participant and a score for each item such that the differences between a participant's score and the item scores have the same order as the participant's ranking of the items, for all of the participants.

So far we have looked at situations involving combining questions, each of which had their own continuum (even if the data could show only discrete points on that continuum and even if the continuum represented only ordinal information). Sometimes, however, data is captured in different forms, from which we still wish to extract measurement values.

For example, it is not uncommon that pairs of objects can be compared. So, we might be able to test a variety of foodstuffs and make preference statements about pairs—'I prefer A to B'—for a collection of such pairs. With ten objects, there are forty-five potential pairs which can be compared. In an ideal world the objects would have a natural rank order (so that, if A is preferred to B, and B is preferred to C, then A is preferred to C), but in the real world this is not always the case. In such situations we can postulate a score for each object and find scores which preserve as many as possible of the stated preferences. The *Bradley–Terry* model produces scores of this kind.

The Bradley–Terry model essentially produces a single continuum and positions objects on that continuum. This principle has been elaborated in a wide variety of ways, especially in psychology and the behavioural sciences. Some of the ways extend from data based on simple pairwise preferences to data based on orderings of several objects, produced by several raters. Other extensions are based on relationships other than simple order. For example, they might be based on measures (or subjective ratings) of the *degree of similarity* between objects. Such methods are sometimes called *unidimensional scaling* or *nonmetric scaling* methods.

The unity of measurement

Most measurement operations are a mix of the representational and pragmatic, although in our discussion we have deliberately focused on illustrations which lie nearer the extremes of the continuum, so as to bring out the core features. Measurements in the physical sciences tend to have a heavier representational aspect, and those in the social and behavioural sciences more pragmatic—though, as we have seen, there are exceptions. The life sciences and medicine take eclectically from a wide range of positions on the representational/pragmatic continuum.

It is worth stressing that measurements reflect only some aspects of the objects being studied. While sticks do have lengths, and people do have intelligence, these single attributes do not capture everything there is to know about sticks or people. Measurement reduces the universe being studied to one or more dimensions, permitting more ready grasp of relationships, more ready inference about what might happen, and more ready control of the world. But it is a simplification. In management contexts, this observation has been expressed in the cautionary remark 'What gets measured gets done', which has been variously attributed. It refers to the dangers of focusing on just a few measurements.

However, while clearly there are dangers in reducing complexity to simplicity, and reducing objects or organisms which could be measured in an infinite number of ways to a mere handful of measurements, there are also enormous gains.

Chapter 3
Measurement in the physical sciences and engineering

A case can be made that the modern world has been largely built on progress in the physical sciences—in physics and chemistry—along with concomitant progress in engineering and medicine. Certainly, these have the longest history of formal methods of quantification. You might have thought, therefore, that measurement concepts and systems in these domains were complete, with little new to be discovered. This, however, would be to misunderstand the alternation between theoretical and experimental progress. The development of new measurement ideas is often fundamental to experimental progress, and advances in experimental discoveries often lead to novel measurement systems and ideas. This is nicely illustrated by the existence of national standards laboratories such as the UK's National Physical Laboratory. This employs over 500 scientists and focuses on the physical sciences, with the objective of 'standardising and verifying instruments, for testing materials, and for the determination of physical constants'.

As well as new physical phenomena needing to be measured, which means that new kinds of instruments have to be devised, even old physical units are regularly redefined. We saw examples of this in Chapter 1, where, instead of defining units in terms of arbitrary etalons, they were defined in terms of natural immutable constants of the universe. We also saw that, as science and engineering progressed, so it became necessary for measurement accuracy to increase. A windmill may have worked very well with tolerances of 1/16th of an inch, but a modern jet engine requires something better.

This chapter looks at measurement of some important physical properties.

Length and distance

Length is one of the most basic of measurements: we can *see* the lengths of objects, whereas the same is not true of weight, temperature, or pressure. Optical illusions aside, one consequence is that it is obvious, when looking at two objects placed side by side, which is the longer of the two. Note that this is even in contrast to area and volume, where, though they can be seen, the shape of the objects complicates the comparison.

Because of its fundamental and easily grasped nature, measurements of a great many other properties are obtained by converting them to length. So, for example, mercury thermometers measure temperature by converting it to the length of a column of mercury in a capillary tube, and weight may be measured by the length to which a spring is stretched, in a spring balance.

There is also a close relationship between length and angle. The tip of the hand of a clock traces out a distance as it moves round the circumference of a clock face, and the needle of a dial traces out a distance as it swings around its pivot. The distance is typically marked out by numbers or graduations, showing how far the tip has moved, but these numbers are also equivalent to the angle through which the hand or needle has rotated.

Note that something of a hidden revolution has occurred in practical measurement in the past few decades, as a consequence of advances in electronics. Instead of giving the results as an angle or length of circumference traced out, instruments now often give numbers directly. This is easier, and probably means that there is less scope for introducing human error (though this can never be completely avoided), but it also has other implications. One is that it has the potential for introducing spurious accuracy. A digital readout to six decimal places may well mean that the last few are subject to substantial uncertainty and should not be taken too seriously: if the room temperature changes, perhaps the last few digits will change, for example. Another is that the hidden transformation from the analogue to the digital disguises the subtleties of measurement and what it means—the sorts of concepts discussed in this book. Attributes do not really come with numbers attached to them indicating their magnitudes.

In Chapter 1, we saw that trade was a key historical driver behind the development of measurement concepts in general, determining quantities of goods of different types. For length in particular, surveying and construction provided similar seminal applications.

An early system for measuring length was by means of a cord which had knots tied on it at regular intervals (so we are back to counting the number of copies of a standard length—the gap between knots). This principle goes back to the ancient Egyptians, and probably also Stonehenge. Note that we can easily construct right angles using such ropes: a closed loop of rope with 12 knots can be arranged as a right-angled triangle, with sides of 3, 4, and 5 gaps between knots. Add a plummet, a weight on a string to determine the vertical, and we have the basic instruments for constructing cathedrals.

In Chapter 1, we explored representational measurement by using length as an example, and we also looked at the history of basic unit lengths. A number of unit sticks laid end-to-end gave the length of an object. Exactly this principle is used in tape measures, or when stretching ropes or chains across a distance. But there are other ways we could determine the length of an object. In particular, we could use an *indirect* measurement procedure. *Tacheometry* is the process of measuring length indirectly.

Ancient indirect ways of measuring length or distance have included how far a stick could be thrown, how far a shout carried, and how far a person could walk between sunrise and sunset—though clearly these do not yield very precise measurements. The example of how far someone can walk in a given time has been generalized to the time it takes something travelling with a known speed to traverse the length. Laser rangefinders use this principle, where the known speed is the speed of light, and radar methods, also involving travel times of electromagnetic signals, use it to determine the distances to other planets.

Another strategy is to use trigonometry. If you know the length of a ruler, then you can work out how far away it is, based on the angle it subtends at your eye (provided it's perpendicular to your line of sight). Conversely, if you know the distance to an object, then trigonometry allows you to work out how long it is from the angle it subtends. A variant of these ideas is that of parallax, which was very important in early astronomical measurements. The effect is illustrated by focusing your eyes on a nearby object and then closing first one eye and then the other: more distant objects will appear to change position, a consequence of the fact that your eyes are looking at the nearby object from different angles. Once again, trigonometric calculations can be used to relate the angle to the distances.

The example of trigonometry illustrates a complementary relationship between the very large and the very small. Increase the accuracy of measuring very small angles, and you can increase the very large distances which can be measured. The Hipparcos space mission used parallax to measure the distance of celestial objects based on accuracies of angles measured in milliarcseconds. A milliarcsecond is one 3,600,000th of a degree.

Other indirect ways to measure lengths include hodometers (or surveyors' wheels; wheels connected to a handle, pushed along by a walking person and where the distance travelled is determined by the number of rotations of the wheel; see Figure 3) and tellurometers (based on measuring the phase shift of light waves, for distances of some kilometres).

What is apparent from these examples is that different procedures are appropriate for different distances. A caliper is appropriate for very small lengths (typically these would be called widths), a foot ruler is appropriate for lengths on a human scale, laser rangefinders, hodometers, or tellurometers for longer scales on Earth, radar for distances to other planets, parallax methods for distances to nearby stars, and so on. The reason different instruments are used is simply practicability. Measuring the distance to a nearby star using end-to-end concatenation of a foot ruler is impracticable to say the least, and measuring a human height using a caliper would be difficult at best.

The need for different procedures becomes even more striking when measuring cosmological distances. We have already noted that the Hipparcos space mission used parallax measurements. For even longer distances other indirect measurement procedures are needed. One is based on the brightness of distant objects. A car headlamp is brighter close up than it is 2 km away, which means that knowing its basic brightness enables us to determine its distance based on its *apparent* brightness. Having said that, as always, things are more complicated than they might seem at first: in astronomy it is necessary to allow for absorption of light by intervening clouds of interstellar dust, for example. Furthermore, it is also necessary to know the object's intrinsic brightness. Fortunately, certain types of astronomical objects, such as supergiants and supernovae, seem to have a relatively standard brightness. This has been determined by measuring the brightness of several such objects at the same distance—where this distance is measured by some other method.

3. A statue of the 18th-century road builder Blind Jack Metcalf, with his surveyor's wheel.

This last example introduces the notion of *metrological traceability*: a chain of distance measures is built, successively suitable for measuring larger distances, and in which each overlaps the previous so that they can be calibrated. By this means scientists have created a ladder of methods for measuring greater and greater distances: the *cosmological distance ladder*.

As we move beyond the simplicity of 'end-to-end concatenation of a standard ruler', the more elaborate measurement methods must be based on various assumptions—for example, that certain types of astronomical objects have a brightness calculable from their properties. If the assumptions are wrong, then the distance measurements are wrong. And in fact that turned out to be the case in one example of astronomical measurement. One type of astronomical object in which brightness is used to determine distance are so-called Cepheid variables. Cepheid variables pulse with a period related to their intrinsic luminosity. However, in the 1950s it was discovered that there were *two* classes of Cepheids, and that the more distant ones belonged to a brighter class than the nearer ones. A consequence of this was that the Milky Way turned out to have twice the diameter previously thought. Once again, the risk of such errors can be minimized by using the notion of convergent measurement procedures mentioned in Chapter 2: if different methods of measuring distance yield the same result, we will have more confidence in the result.

Area and volume

In Chapter 1, we saw how early units of area were defined in terms of the time land would take to plough, or how much volume or weight of crop it would produce. Other early methods define area in terms of the length of the boundary—for example, how long the land would take to walk round. This, of course, can cause difficulties, as is illustrated by the tale of Dido in the 8th century BC. Legend has it that King Hiarbas, near Carthage, agreed that Dido could buy as much land as she could 'encompass' with a bull's hide. Dido proceeded to cut the hide into very thin strips, with which she surrounded a large area. The fundamental point—and the difficulty of area measurement—is that the area is dependent on the *shape*: a misunderstanding by Dido and the same length of strips of leather would have enclosed a very long thin shape with a very small area.

In principle, we could use a basic representational method for measuring area. Define a unit area—say a very small square—and then measure a larger area by covering it with copies of the small square, not overlapping and leaving no gaps of course. That's fine provided the larger area has a nicely rectangular shape, but problems can arise if the edge of the larger shape is diagonal or curved: then it is not possible to leave no gaps or have no small squares projecting beyond the edge of the large area. The problem is eased by reducing the size of the small square (an idea which ultimately leads to the mathematical concept of integration).

Measuring an area of land used for agricultural purposes is one thing, but measuring other areas can lead to different approaches. For example, the symmetry of (circular) pipes means that the cross-sectional areas can be easily determined from their diameter—which can be measured using calipers (for small pipes) or standard rulers or tape measures for larger pipes. Alternatively, you can wrap a tape measure around a pipe to determine its circumference, and convert that to a cross-sectional area. There even exist tape measures where the conversion has been done for you, so, for example, instead of the markings on the tape showing a circumference of d inches, they will show a cross-sectional area of $\pi d^2/4$.

The relationship between length and area may be obvious to us nowadays: the area of a rectangle is simply its length times its breadth (both being length measurements). But this familiarity is a consequence of long usage, and may not be so obvious a priori. This point also applies to volume, where, like area, shapes also matter.

To measure volume, we could start with a tall thin container with regular cross-section and make calibration marks at equal distances along it, as in a burette: we are measuring volume in terms of length. Or we could take a basic unit (a small cup say) and count how many cupfuls it takes to fill a container with water. This allows us to measure the internal volume of containers in terms of number of unit cups—which is critical for trade and other activities—but it does not allow us to measure external volume. As Archimedes showed, a simple way to measure external volume, such as the volume of a gold crown, is to immerse it in water in a calibrated container: the volume of water displaced, as read from calibrations, is equal to the volume of the object.

Mass, weight, and force

A basic tool for measuring weight is the balance or scales, consisting of an arm pivoted about a fulcrum at its centre, with pans suspended from the two ends. Objects placed on one pan are balanced against objects placed on the other. We can then use a basic representational approach to construct a system of measurements, balancing new objects on one side against multiple copies of an object of unit weight on the other.

This approach, based on the idea of the lever, has been extended in a variety of ways. The steelyard is a version which has unequal length arms (so a weight of one unit on one side balances a larger weight on the other). The bismar is a version which has a fixed weight at one end, and balances the weight at the other end by moving the fulcrum.

A different strategy is taken in the spring balance, mentioned in Chapter 2. This is a device based on Hooke's Law, which says that the extension of a spring is proportional to the applied load or force. In a similar way, other physical properties have been used to map weight to numbers—or to length and thence to numbers. They include the torsion balance, the flexure strip balance, and the oscillating quartz crystal balance.

Weight is not the same as mass. Weight is the gravitational force exerted on an object by the Earth. Mass is the amount of matter in an object. However, at any particular point on the Earth the attraction from the Earth will remain constant, so the mass of objects is proportional to their weight at that point. This means that we can use any of the measuring instruments just discussed to determine mass. However, were we to take an object from the Earth to the Moon, we would find that those instruments based on gravitational attraction (e.g. a spring balance) would yield a different weight for the object: they would need recalibrating to determine the extension caused by a unit mass. Weighing scales will still do the job though, since the change in attraction when we move from the Earth to the Moon will be the same for both sides of the scales: an object will be balanced by the same number of copies of a unit mass whether the measurement is taken on the Earth or the Moon. The SI unit of mass is the kilogram, described in Chapter 1.

Newton's second law tells us that when we apply a given force to an object its acceleration is inversely proportional to its mass. So this is another (and an indirect) way to determine mass: apply a fixed force, and see how fast the object accelerates. Conversely, of course, we can measure force by seeing how fast an object of a given mass is accelerated when the force is applied to it. The SI unit of force is the newton—equal to the force needed to accelerate a kilogram mass at 1 metre per second per second.

When a force causes an object to change position (e.g. lifting something into the air, or a voltage causing electric charge to flow), *work* is said to be done, and obviously the amount of work done can be measured. The SI unit of work is the joule (after James Prescott Joule).

Work can be done quickly or slowly: you can lift a rock with a sudden jerk or in a gradual movement. The rate at which work is done is called *power*, with an SI unit of the watt (after James Watt). An older unit was the horsepower, equal to about 750 watts.

The decibel is a useful unit used to express the ratio of the value of a physical quantity, such as power, to a reference value. It is defined as $10 \times \log_{10}R$, where R is the ratio of the two values. The use of the logarithm means that huge scales can be covered without writing huge numbers: a power ratio of 1 corresponds to 0 decibels, a ratio of 10 to 10 decibels, and a ratio of 100,000 to 50 decibels.

Time

We have seen the advantages of abandoning idiosyncratic units of measurement, but in one area such a simplification has not yet been made: the measurement of time. We still have seconds, minutes, hours, days, and weeks, with respective aggregating factors of 60, 60, 24, and 7. Over the span of history various attempts have been made to rationalize our present system—such as a French attempt to change to days consisting of 10 hours, each with 100 minutes, each with 100 seconds. But the present system has stuck, with the exception that larger and smaller units, those beyond everyday human experience, are defined in powers of ten. Thus we have millennia, decades, milliseconds, picoseconds (1 picosecond is 10^{-12} seconds), and so on.

Many of the units of time have their origins in planetary astronomical phenomena—where the natural periodicity (at least, roughly) led to natural (and convenient) time intervals. It is this periodicity and regularity which is the key to measuring time—where the period defines the basic unit—and many other regular physical phenomena have also been adopted as the basis of clocks. They include familiar ones such as pendulums and dripping water, but also more exotic ones such as vibrating quartz crystals, caesium atoms, and the frequency of electromagnetic waves. Most recently, so-called 'optical lattice clocks' have been developed, in which strontium atoms switch between energy levels with a precise frequency when they are illuminated with red lasers. Molecular and atomic systems have the advantage that they do not become fatigued, fractured, or wear out.

Just as measures of length, weight, and volume were critical for the development of civilization in such things as trade, so also were measures of time. In Chapter 1 we mentioned the well-known example of measuring longitude, for which an accurate clock was critical, but a more immediate example would be the time for which perishable food could be stored.

Accurate time measurement, especially of very small intervals, is important in much scientific research. The most accurate clocks so far developed have an error of about one second in 16 billion years. They are so precise that they can detect slight changes in the gravitational field of the Earth: the theory of general relativity tells us that clocks run slower in stronger gravitational fields.

As with many other scientific units, the definition of units of time has changed as advances have been made. The second's old definition as 1/86,400 part of a mean solar day has undergone several changes, to the modern definition of the duration of 9,192,631,770 periods of the radiation corresponding to the transition between the two hyperfine levels of the ground state of a caesium 133 atom at rest at a temperature of 0 kelvin (kelvins are defined in the section entitled 'Temperature'). Once again we see the move from measurement definitions in terms of macroscopic physical objects, subject to all the wear and decay associated with such things, to fundamental physical properties, which are immune to them.

Temperature

Just as we can feel weight, so we can feel temperature, to the extent that we can say that one object seems warmer than another.

However, while it is obvious that temperatures (at least for objects of the same material) can be ordered, albeit perhaps not with very fine distinctions, it is perhaps not so obvious that they permit a higher level scale—an interval or ratio scale. This is why W. E. Knowles Middleton called early thermometers thermo*scopes*, for *seeing* rather than *measuring*, and for producing results on a purely ordinal scale (Galileo is sometimes claimed to be the first inventor of such a device).

Whereas we can concatenate multiple copies of an object with a particular length to see how many are needed to stretch as far as some other object, we cannot concatenate multiple copies of an object with a particular temperature to determine the temperature of another object. Temperature is sometimes termed an *intensive* measurement: put two objects (say two quantities of a liquid) of equal temperature together and you get another (larger) object of the *same* temperature, not twice the temperature. Put two objects of different temperatures together and (once things have settled down) you get something with an intermediate temperature—and, worse, a temperature which depends on the mass and material of the objects.

While we might define temperature as 'degree of hotness', its true nature is not immediately obvious. The story of the elucidation of temperature is the story of the understanding of a fundamental aspect of physical science: the nature of energy. Temperature also provides perhaps the most obvious illustration of indirect measurement. Apart from ordering temperatures by subjective sensation, all temperature measurement is in terms of the effect different temperatures have on other physical properties: the change in volume of fixed quantities of mercury or alcohol, the speed of sound in a substance heated to different temperatures, the character of light radiated from heated materials, the electrical resistance or conductance of a heated substance, the expansion of a metal, the pressure in a fixed volume of gas, and so on.

The historical starting point in establishing numerical scales for temperature is to determine some fixed points, and a wide variety have been suggested. Newton proposed the freezing point of water, the temperature at which wax melts, the boiling point of water, the temperature of burning soft coal when fanned by a bellows, and others. The Danish astronomer Ole Roemer suggested using the temperature of the coldest mixture which could be attained by mixing salt, ice, and water, the temperature of freezing water, the temperature of the human body, and the temperature of boiling water as fixed points. The use of the human body here is reminiscent of defining early length measurements in terms of human dimensions. Other proposed fixed points included the temperature of candle flames, snow, the solidifying point of aniseed oil, and the temperature in deep caves. Daniel Gabriel Fahrenheit based his scale on Roemer's, but divided each degree into four and took as the zero point the temperature obtained when ammonium chloride was added to the salt, ice, and water mixture—which is why the Fahrenheit scale has a temperature of 32° for the freezing point of water. Anders Celsius, after whom the Celsius temperature scale is named, originally chose 0° for the boiling point of water and 100° for the freezing point, inverted relative to today's scale. For obvious reasons, this scale was originally called the Centigrade scale. You will probably have spotted that the adjective 'fixed' in all this discussion needs to be interpreted with a pinch of salt.

Once at least two fixed points have been chosen, they can be used with one of the physical properties we have described to produce a graduated temperature scale. This is done by dividing the range of change of the quantity (the length of a mercury column, for example) into equal intervals. Note, however, the fundamental assumption behind this: that equal changes in temperature lead to equal changes in the magnitude of the physical quantity. In fact, this assumption is generally unfounded, so that different physical properties will lead to different scales: a unit length increase in the length of a column of mercury may not correspond to a unit change in angle resulting from a coiled bimetallic strip, for example. This means that temperature scales defined in such ways are heavily pragmatic.

Defining temperature scales for temperatures above or below the fixed points mentioned above requires new fixed points. Examples which have been used include, at one extreme, the melting point of tungsten and other metals, and, at the other, the boiling point of oxygen and the triple point of hydrogen. But defining the scale is not enough: at high enough temperatures mercury will vaporize and glass will melt, so other methods are needed. Pyrometers are used to measure very high temperatures.

Newton developed a method based on how long it took a very hot object to cool down to a known temperature. The potter Josiah Wedgwood tackled the problem of measuring temperatures inside kilns, where metals melt, using a method based on the contraction of small pieces of clay when they were heated. He specified the size of the pieces (cuboid shapes, 0.6 by 0.4 by 1 inch) and described how to determine the temperature using a numerical scale on a brass gauge. He linked the numerical results from his *contraction pyrometer* to the standard Fahrenheit scale using the expansion of silver with heat, since this overlapped the two other procedures—illustrating again the use of a ladder of measurements.

The list of temperature-dependent physical phenomena which have been used to develop pyrometric measurements is endless. More recently they include the characteristics of electromagnetic radiation emitted by a hot object, and this brings us closer to a representational notion for temperature.

The pragmatic arbitrariness of the temperature scales produced by these approaches will be obvious, but latent within some of them is the hint that there might exist some more fundamental (and less pragmatic) approach to temperature measurement. This is the observation that the pressure of a fixed volume of gas decreases as the temperature is lowered. Extrapolation down to zero pressure suggested that there might be a minimum possible temperature—an *absolute zero*. Using this notion, we can redefine standard scales like the Fahrenheit and Celsius scales, so that they start at this absolute zero (instead of at the temperature of an ammonium chloride/salt/ice/water mix, or the freezing point of water, etc.). When this is done with the Celsius scale, we obtain the *kelvin* scale, with units denoted K (note, not 'degrees K'). The freezing point of pure water under normal atmospheric pressure is 273.15 K.

At this point we have described a measurement scale which has a zero point chosen using representational considerations, but with a pragmatic choice for the units. Indeed, we have not yet uniquely defined a representational unit of temperature, because, as we noted, different physical properties may yield non-linearly related scales. This means that if we have two thermometers based on different physical phenomena, even if both are graduated to show 0 degrees at the freezing point of water and 100 degrees at the boiling point, and both have the intermediate range divided into 100 equal intervals, they may well show different temperatures when plunged into the same warm body of water. To disentangle things, we need to dig more deeply into exactly what temperature is.

Matter is composed of atoms and molecules in constant motion—perhaps vibrating, as in solids, or moving along trajectories, as in gases and liquids. Since they are in motion, they have kinetic energy. A 'perfect gas' is an ideal (and hence imaginary) material whose particles are so small that they are mere points which do not interact with each other except when they collide. That means that the particles would travel in straight lines, with constant velocity and hence constant energy between collisions. The average energy of the particles is the *thermodynamic temperature*.

The pressure due to a gas arises from the bombardment of the walls of the container by the particles composing the gas. Clearly, the faster the particles are moving (that is, the greater the energy they have), the greater the pressure.

Pressure and thermodynamic temperature are each proportional to the average kinetic energy of the particles, so that doubling the thermodynamic temperature would mean that the pressure doubled. In particular, this means that we could use the pressure exerted by an ideal gas as a thermometer—and the pressure would be zero at a temperature of absolute zero. Of course, since a perfect gas is an idealized fiction, we can only approximate it in practice, but helium provides a good approximation.

If absolute zero provides one fixed point for our temperature scale, we need another to be able to define units. Nowadays the *triple point* of water is used to define the basic SI unit of temperature, the kelvin. The triple point is the temperature and pressure at which water exists in equilibrium in its gaseous, liquid, and solid form, and it occurs at 273.16 K, so that the kelvin is defined as 1/273.16 of the temperature of the triple point of water. Note that all of this development of an understanding of heat as a form of energy, and its relation to temperature, entropy, and other concepts occurred late in human history—long after pragmatic temperature scales had been developed and used. This parallels the use of notions of length millennia before sound formal representational theory for length measurement was developed. And it also illustrates the interplay between growing understanding and advancing measurement technology.

Electrical and magnetic units

We have seen something of the vast number of different units of measurement coined for length and weight. Although the understanding of electricity and magnetism has a far shorter history than those two physical concepts, these have also generated a very large number of units. The reason is partly because notions of electricity and magnetism span a variety of concepts, and partly because it took substantial effort by many researchers to tease them out.

The measurement of electrical and magnetic phenomena has been driven by a number of practical applications, especially in engineering. We have already seen that applications often require accurate measurements, and here we find the needs of the telegraph, telephone, electric motors, heaters, light, radio waves, and a host of other electrical phenomena influencing the development of measurement technology.

We can begin with the notion of electrical charge—an *amount* of electricity. Early recognition that there was such a thing can be traced back several centuries BC, with the observation that amber rubbed on rabbit fur attracted light objects, and even produced sparks. Some 2,000 years later investigators developed mechanical devices which could generate and store this electric charge. The idea of 'storing' something hints at the notion of 'quantity', though it is far from indicating how that quantity might be determined.

As with temperature, early attempts to measure electrical charge led to results on an ordinal scale. The electroscope, for example, consists of a vertical metal rod with two parallel strips of very thin gold leaf at the end. A charged object brought to the top end of the rod causes the gold leaves to become charged, and since like charges repel, the leaves bend apart from each other. More elaborate electrometers are based on the same principle. For example, two balls of pith, one fixed and one suspended from a glass thread, may be made to repel one another by connecting them to a charged object, so that the suspended ball moves through an angle which depends on the charge.

The measurement of different aspects of electricity also provides many examples of indirect measurement. For example, voltage is typically measured by seeing how much current flows through a given resistance, and relating voltage and current using Ohm's Law. And charge may be measured in terms of voltage. Such indirect measurement also permits very small and very large charges to be measured—just as developments in pyrometry extended the range of temperatures which could be measured.

The measurement of electrical charge is a nice illustration of how progress in scientific understanding and progress in measurement technology support each other, with advances in the one leading to advances in the other, and vice versa. In particular, it was discovered in the early 20th century that electrical charge comes in multiples of a very small basic unit—the charge on a single electron. In fact this is so small as to be useless for most applications, and instead we use the coulomb, a charge approximately equal to that of 6.24×10^{18} electrons, named after the French physicist Charles-Augustin de Coulomb. The coulomb is the SI unit for charge.

The *flow* of electrical charge—that is, electrical current—can also be measured in various ways. The basic SI unit of electrical current is the ampere (named after André-Marie Ampère, and often abbreviated to 'amp'), defined as a flow rate of one coulomb per second. For many practical purposes, the ampere is much too large, and the milliamp is used.

Electrical charge will flow when there is a potential forcing it to flow, and the measure of this electrical potential is the voltage. One volt (named after Alessandro Volta) is the difference in electrical potential between two parallel infinite plates 1 metre apart that creates a force of 1 newton per coulomb between the plates.

And so it goes on, with units for resistance (the ohm, after Georg Ohm), capacitance (the farad, after Michael Faraday), inductance (the henry, after Joseph Henry), magnetic flux (the weber, after Wilhelm Eduard Weber), magnetic flux density (with both the tesla, after Nikola Tesla, and the gauss, after Carl Friedrich Gauss), frequency (the hertz, after Heinrich Hertz), and so on.

Apart from the sheer number of concepts needing to be defined and measured, things are aggravated by the fact that different systems of units had different names attached to them (think back to the varieties of units and names for length), and also the fact that the definitions change over time as a more solid theoretical understanding has been developed, leading to more sophisticated indirect measurement procedures being devised.

Quantum measurement

This chapter would be incomplete without a mention of the role of measurement in quantum theory. This theory describes the behaviour of the universe at very small scales. These are scales completely beyond our everyday experience, so it should perhaps not be surprising that behaviour at these scales often runs counter to our intuitions. After all, if the air has the consistency of treacle to a small flying insect, how much more strange should we expect things to seem at the scale of subatomic particles?

Quantum mechanics tells us that certain pairs of properties cannot simultaneously be measured with arbitrary accuracy. This is not because the measurement process necessarily disturbs the object being measured (even though this is also true—at a minimum, to measure something we must bounce a photon off it, and this will have an effect). Rather it is due to the fundamental nature of the universe. An example is given by the position and momentum of a small particle. We can measure the position as accurately as we like, but the relationship between the two means that the accuracy is gained in terms of a cost of inaccuracy in knowing the momentum. And vice-versa. Furthermore, it turns out that it is not meaningful to speak of the true value of an attribute prior to measuring it. Rather, the very action of measurement 'collapses' the attribute to a particular value. This is clearly very different from classical physics, where the properties of objects have values, simply waiting to be measured. While the world of quantum mechanics is a truly strange one from the perspective of classical physics, countless experiments, and indeed machines built using the theory, show that it is essentially correct. It may be strange, but when tested in experiments its predictions are right.

And beyond

This is a short book, so in this chapter it has been possible only to touch on the vast array of measurement methods and units which have been devised for the physical sciences. Units for more than a hundred physical attributes have been described, and entire books have been written on many of them. A curious phenomenon, however, is that (with some very few exceptions) physical properties can all be expressed as monomial combinations of six base attributes. If we choose charge (C), temperature (R), mass (M), length (L), time (T), and angle (A) as the six base attributes then: speed is length (= distance) per unit time, or LT^{-1}; density is ML^{-3}; electric current is CT^{-1}; magnetic flux density is $MT^{-1}C^{-1}$; and so on. This has some rather useful implications, to which we return in Chapter 7. In particular, it means that you can check whether the equations of proposed theories could possibly be right by looking to see if they have the same units on both sides of the equals sign.

Chapter 4

Measurement in the life sciences, medicine, and health

Biological systems pose particularly challenging measurement issues. This is partly because the biological domain is characterized by complexity and diversity. Biological systems (even something so basic as a cell or bacterium) typically have a considerable internal complexity, which can take many different forms, so generating a huge number of different kinds of organism. Moreover, biological systems interact in complicated ways with their external environment. This means that they are constantly changing, so that even defining the attribute to be measured can be difficult.

As a consequence, the results of measuring biological organisms will often result in a *distribution* of values. A familiar example of this is the distribution of weights or heights of the people in a human population. Less obviously, it means that the value of some indicator (say a hormone level) which is perfectly reasonable and normal for one person may be abnormal and even dangerous for another, perhaps indicating a pathological condition.

A further complication in medicine is that much measurement is of attributes or characteristics relating to internal or subjective phenomena—of pain, anxiety, dizziness, and so on. This means that, in a medical context, it may be necessary to rely on subjective reports from a patient. I stress that the word 'subjective' here refers to the subjectivity of the patient, not that of the reporting physician, even though his or her determinations of severity or magnitude may also be subjective (e.g. in interpreting X-ray or ultrasound images).

With considerable justification, we might expect measurements based on subjective determinations to be less reliable and generally to have poorer measurement characteristics than those based on sound representational relationships. However, that very difficulty has meant that there has been considerable research effort expended on measuring subjective phenomena, so that sound procedures do exist. In medicine, an objective measurement, something that can be observed or detected by someone else, is called a *sign*. In contrast, a subjective measurement identified by the patient as indicating abnormal functioning is called a *symptom*.

I have remarked on the resistance to the introduction of measurement concepts. Medicine is one of the domains in which this resistance was most pronounced—with the proponents and the opponents fluctuating in terms of who had the upper hand.

We can trace the use of measurements in drug prescriptions at least back to 1500 BC, in ancient Egypt. But the intellectual leap from the familiar physical measurements such as volume or weight of drugs to measurement of aspects of patients is a big one. Isolated examples occurred, such as Hippocrates measuring the pulse, but its more general adoption had to wait some thousands of years. Even as recently as 1797, the work of James Currie on medical measurement of temperature was used by a German translator as an example of the backwardness of English medicine. The general perception seemed to be that measurement would give numbers, but would not yield the 'qualities' deemed necessary for medicine. It was only around the middle of the 19th century that pulse rates, body temperature, blood proteins, blood pressure (see Figure 4), and so on began to

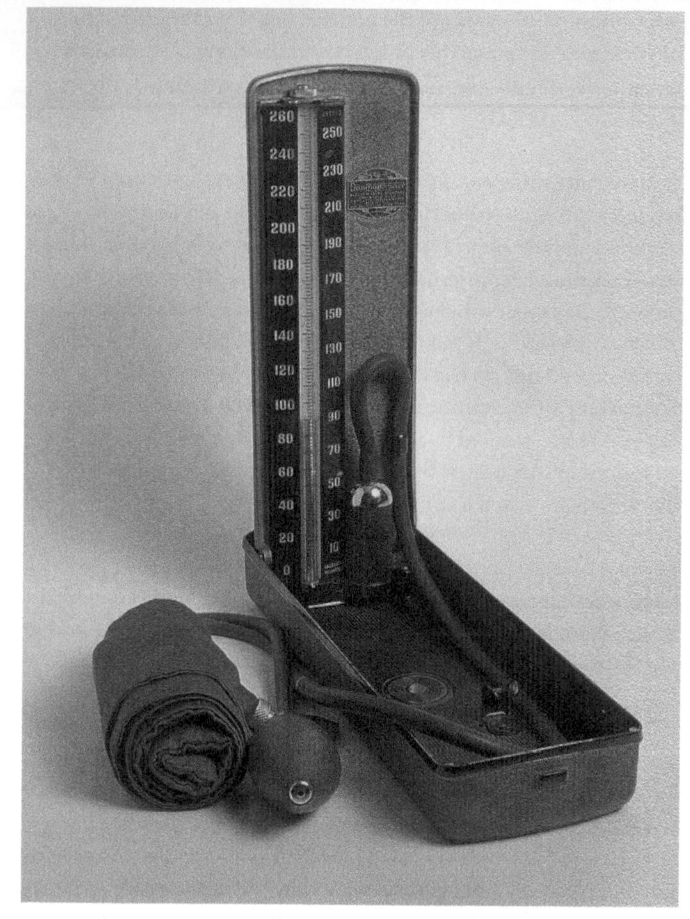

4. A mercury manometer, for measuring blood pressure.

be regularly used, even if they had been described much earlier. Also around this time, other instruments, such as the stethoscope, laryngoscope, microscope, and opthalmoscope, began to be used (and note the *scope*, as in our discussion of thermoscopes and electroscopes in Chapter 3).

These instruments built on the demonstrable success of numerical methods applied in epidemiological contexts, which could not be dismissed. 'Bills of mortality', showing death rates from different causes, and permitting different social groups to be contrasted, had provided clear examples of the success of measurement, albeit at the population level. Of course, these successes were not without their difficulties, and finer use of such epidemiological data was held back by a lack of clarity about the nature of the classifications. What should we make of 'convulsions', 'decay', or 'fevers' as diagnostic categories?

Another source of resistance to these statistical demonstrations was that they were just that—statistical. They applied en masse, to averages, and not necessarily to the individual case. Nonetheless, they gained ground, and also proved effective in more focused studies. Pierre Louis, for example, had attacked informal comparisons and had illustrated numerical precision in a comparison of the efficacy of bloodletting in 100 patients with pneumonia by contrasting the outcome with a control group. In 1840 Henry Holland described medical statistics as the most secure path into the 'philosophy of medicine', and in 1855 W. P. Allinson said that there was 'no other way' to investigate many of the most important questions in medicine than through statistics.

Scope of medical measurement

In modern medicine, there are many different reasons for taking a measurement, and hence many different types of measurement. Measurements are taken to arrive at a diagnosis, to determine prognosis, to decide on disease stages, on severity, health-related quality of life, appropriate treatment and dose regimes, effectiveness of treatment, and for other reasons. Different measurement procedures might be appropriate, even for the same attribute, if the purpose is different: measurements aimed at diagnosis need to discriminate between conditions, while measurements aimed at evaluating the effectiveness of treatments will need to be sensitive to changes in conditions.

Examples of the different purposes, and hence different needs in measuring instruments, are given by 'activities of daily living' (ADL) studies (measuring the capacity people have for undertaking normal tasks), where we might be interested in what patients *say* they can do, what they are *able* to do, and what they actually *do* do. Clearly some sort of interview or self-completed questionnaire may suffice for the first, a task-based test will be needed for the second, and an observational test for the third. Another example is given by Alvan Feinstein in *Clinimetrics*, his classic work on medical measurement. He points out that the approach to measuring congestive heart failure should depend on whether the measurement is to be used to detect failure, measure its magnitude, predict the likely outcome, or to decide how to treat it.

One important distinction is between basic measurements of bodily condition and function on the one hand, and higher level indicators of an adverse medical condition on the other. The dividing line is not always clear, but it is a useful taxonomic classification for discussing measures.

Another distinction is between measurements which are direct observations of a unidimensional concept, and those which are based on summarizing several other measures. Typically more basic or lower level indicators will be unidimensional—measuring, for example, temperature, concentration, volume, area, pain intensity, blood pressure, and so on—whereas higher level measurements will often be formed by combining multiple measurements. And here we should recall the distinction between clinimetric and psychometric approaches, discussed in Chapter 2. The core of this distinction lies in whether we are seeking to combine a variety of diverse indicators to give an overall measure, or seeking to extract a measure of some common underlying influence from a collection of indicators.

If statistical tools are being used to construct the measures by combining several, more elementary, indicators, then different, albeit related, kinds of statistical methods will be needed for different purposes. For diagnosis, for example, tools related to statistical discriminant analysis will be needed, aimed at constructing measures which distinguish between different conditions. On the other hand, for evaluating the effectiveness of treatments, statistical methods for detecting change will provide the answer. Tools for prognosis will be based on statistical methods of prediction, possibly modelling the known course of a disease, or possibly purely empirical models, based on observed outcomes of previous patients.

As with all other measurement procedures, medical measurements sit on a representational/pragmatic continuum. Pre-clinical or laboratory measurements of physiological systems often tend towards the representational. Temperature, blood pressure, pulse rate, erythrocyte sedimentation rate, and blood glucose level are general examples, and different bodily systems will often have associated collections of tests. For example, kidney function is measured by tests which include assessment of levels of urea, creatinine, glomerular filtration rate, and electrolytes.

On the other hand, although a raw measurement may be strongly representational, in many cases it will typically be used as an indicator of some underlying condition, so that it might better be regarded as pragmatic. For example, tumour size measured by the length in millimetres of its maximum diameter is clearly a representational measure, with all the properties that implies. But here length is being used as an indicator of severity. Moreover, unlike the mapping from temperature to length implicit in a mercury thermometer, the mapping here is much less well defined. 'Severity' is not such a clear concept. This means that the size of a tumour, measured using length, is better thought of as a pragmatic measurement—certainly doubling the size of the tumour does not mean twice the severity.

In contrast to pre-clinical measurements, clinical measurements, those taken in clinical examinations, indicating severity or degree of some clinical condition or manifestation, are more heavily pragmatic. A simple example is given by cancer stage systems. One such system is a pragmatic scale defined as:

> Stage 1: the cancer is relatively small and contained within the originating organ;
>
> Stage 2: the cancer is larger than in stage 1, but has not spread into surrounding tissue;
>
> Stage 3: the cancer is larger, there are cancer cells in the lymph nodes, and it may have spread to surrounding tissues;
>
> Stage 4: the cancer has spread to other organs.

In a similar vein the New York Heart Association (NYHA) Functional Classification of heart failure is a pragmatic scale with four categories:

1. Patients with cardiac disease but without resulting limitation of physical activity. Ordinary physical activity does not cause undue fatigue, palpitation, dyspnea, or anginal pain.
2. Patients with cardiac disease resulting in slight limitation of physical activity. They are comfortable at rest. Ordinary physical activity results in fatigue, palpitation, dyspnea, or anginal pain.
3. Patients with cardiac disease resulting in marked limitation of physical activity. They are comfortable at rest. Less than ordinary activity causes fatigue, palpitation, dyspnea, or anginal pain.
4. Patients with cardiac disease resulting in inability to carry on any physical activity without discomfort. Symptoms of heart failure or the anginal syndrome may be present even at rest. If any physical activity is undertaken, discomfort is increased.

Implicit within these examples are multiple lower level measurements being combined into a higher level pragmatic scale. In general, observations on several physiological processes, or observations of symptom constellations may be combined (some psychiatric conditions are defined in this way, for example). When this happens, as we saw in Chapter 2, several questions need to be answered: what variables to choose, how to score them, whether to transform them, and how to combine them.

Different measurement cultures have different answers to these questions. In medicine, the strategy has mainly been to identify relevant aspects based on prior or historical clinical experience (recall Virginia Apgar's description: 'A list was made of all the objective signs which pertained in any way to the condition of the infant at birth'). In contrast, in the social and behavioural sciences, the strategy has typically been to collect a large number of potentially relevant items (or questions), and then reduce this set to a critical core set by a combination of statistical and expert panel evaluations.

The driver behind pragmatic measurement is that it should yield something useful. With this in mind, in measures constructed by combining multiple lower level indicators there is a tension between including a great many such indicators (better coverage of the content, reduced variability through aggregation) and including just a few (easier to administer, more likely to obtain responses, possibly more ready interpretation).

How to measure patients

Many medical measurement procedures are based on integrating responses to a series of questions asked of the patient. Broadly speaking, these can be self-administered or administered by a clinician, perhaps as a structured or semi-structured interview. Depression measurement provides many examples of both types. The Beck Depression Inventory (BDI) consists of twenty-one items, each consisting of four or five alternative statements ranging in grades of severity from 0 to 3, with the patient required to choose the statement most closely matching his or her condition. The total score is the sum of the grades. The (usual form of the) Hamilton Depression Scale (HDS) consists of seventeen items covering the symptoms of depression, nine of which are rated on a five-point scale, with the other eight being rated on a three-point scale. Depression severity is the sum of these ratings, and the overall HDS score is the sum of two independent ratings, one obtained by an interviewer and the other by an observer. The subtlety of the structures and grading systems of these instruments is an illustration of the amount of research effort which has gone into producing reliable and accurate measuring instruments for subjective phenomena. These two instruments have also undergone a massive amount of evaluation work over the years since they were first defined (BDI in 1961; HDS in 1967), so their properties and behaviour are well understood.

Pain measurement is another area for which a great many instruments have been developed, spanning self-report, observational, and physiological measures. The challenges are illustrated by the differences in measuring pain experiences for adults, children, and infants, who clearly have different capacities for describing subjective phenomena. While intensity is perhaps the most important dimension of pain, it is not the only dimension—we listed some others in Chapter 2. Adjectives used in the classic McGill Pain Questionnaire, developed in the 1970s, illustrate the range of pain sensations: flickering, pulsing, quivering, throbbing, beating, pounding, jumping, flashing, shooting, pricking, boring, drilling, stabbing, and so on, for a total of seventy-seven adjectives describing different kinds of pain experience, divided into categories, which are combined in an elaborate way to yield an overall score.

A third area, related to both depression and pain, is wellbeing and quality of life. This is a topic which spans a huge breadth. In a health context, and for individuals, however, a common starting point is the World Health Organization's definition of health: 'a state of complete physical, mental, and social well-being, and not merely the absence of disease'. In medical contexts, discussions are often restricted to health-related quality of life (HRQoL). Even with this restriction, however, it is a potentially vast space spanning, as Peter Fayers and David Machin put it, 'general health, physical functioning, physical symptoms and toxicity, emotional functioning, cognitive functioning, role functioning, social well-being and functioning, sexual functioning, and existential issues'.

HRQoL measurement illustrates the importance of having good and relevant measurement procedures in domains where the procedure is necessarily heavily pragmatic: the quality-adjusted life year (QALY) is a unit for the number of extra years of life that would be added by a medical intervention, but where each extra year is adjusted to take the quality of life into account. An extra year in perfect health would count as 1, while an extra year in less than full health would count less.

We have already encountered activities of daily living (ADL) scales, which are used to assess levels of disability and the extent to which an individual can undertake normal daily activities such as feeding, washing, and dressing themselves. As should be expected with heavily pragmatic measurements in social and behavioural contexts, there are many different ADL scales. They will be aimed at different populations (perhaps different age groups, different social contexts, etc.), with different conditions (perhaps people with chronic illnesses, or mobility limiting conditions), and used for different purposes (perhaps how well someone can function in a community, how much nursing care is needed, and so on). As always with such measures, while it is easy to construct a list of questions, score each of them from 0 to 10, and add up the results, doing this in a way which properly taps the underlying concept to be measured and which has desirable measurement properties is a non-trivial exercise. Moreover, some scales require fairly extensive training before they can be used effectively. This serves to standardize those using the scale, and results in greater reliability and, in general, more trustworthy conclusions.

To make medical measurements useful, so that anyone anywhere will interpret them in the same way, the scales themselves need to be standardized. The issue is very similar to that of using the same basic units of measurement—of length and weight for example—discussed in other chapters. In medicine, however, the range of different types of measurement scale means more effort must be put into standardization. For example, it might mean rescaling scores so that a population distribution has some standard form (e.g. a Gaussian distribution with a mean of 100 and a standard deviation of 15), so that extent of deviation from the norm can be readily assessed. In purely pragmatic scales the standardization of the scale is part of the definition of the measurement.

Another issue for those medical measurements which involve asking a patient to give a value is how best to approach this elicitation. We can ask for the information on a discrete scale (e.g. 'Would you say you are much worse, slightly worse, same, slightly improved, much improved?'), on a visual analogue scale (e.g. 'indicate on this line stretching from 0 to 10 how severe your pain is, where 0 is no pain and 10 is the worst pain imaginable'), semantic differential scales (in which a patient is asked to place themselves on a scale ranging between two adjectives with opposite meaning), as well as other alternatives, such as a series of pictures (e.g. a series of cartoon faces with different expressions, from very sad to very happy).

Elicitation obtains information, but there is a complementary aspect: the communication of measurement results. This can also be done in various ways: using numbers, symbols on a graph (e.g. a bar chart), trend lines, and so on. What is most useful will depend on the objectives of the communication.

Although most of this description has been couched in terms of producing a single overall score or measure, when multiple items are combined it is very often possible to produce subscores, rating particular aspects of the overall characteristic. Thus, for example, the Sickness Impact Profile from the Rehabilitation Institute of Chicago is a score of quality of life and level of dysfunction based on sixty-eight questions. However, implicit in the overall score are evaluations of behaviour, life participation, mental health, and social relationships.

Measurement issues in medicine are not restricted to the health of patients. They also arise in other contexts. For example, it is critical that doses of medicines are properly measured. The consequences of failing to do so can be catastrophic: newborn Alyssa Shinn died in November 2006 when, instead of receiving a dose of 330 micrograms of zinc to boost her metabolism, she received a dose of 330 milligrams. The standard 'unit-dose' drug dispensing system uses pre-packaging of unit doses so that they are ready to administer to the patient, meaning that a nurse or other clinician need not measure up the dose on the fly.

Statistical measurements

Returning to the point made at the start of this chapter, biological organisms within a population are not identical. One implication of this is that some will be more resistant than others to damage—and to levels of dose or poison. This fact has been used to define a measure of dose of damaging substance, radiation, etc. The *median lethal dose* (LD_{50}) is the dose needed to kill half the members of a population. The lower the level, the more dangerous the material. Generalizations of this concept have led, through the statistical technique of bioassay, to similar measures—LD_5 is the dose required to kill 5 per cent of the population.

I mentioned at the start of this chapter that one of the early successes of measurement in health was when it was applied at the level of a population rather than an individual. This is where measurement becomes entwined with statistics. To measure a characteristic of a population (e.g. a rate of infection, the proportion of those of working age to those above working age, or the death rate) it is necessary to take measurements on the individuals within that population and aggregate them in some way. We discuss these sorts of issues further in Chapter 6, and here merely note that epidemiology is the study of diseases in populations. Such matters have a long history. Florence Nightingale became a national hero for her measurements and statistical descriptions of illness and disease in populations, such as her demonstration that contaminated water and overcrowding led to high death rates in the British army in India.

Chapter 5
Measurement in the behavioural sciences

Concepts of measurement in psychology are particularly noteworthy for having encountered scepticism. While people have been happy to accept that psychological attributes can be compared ('I like *this* more than *that*') many are suspicious about the possibility of assigning numerical scores to such concepts. As far back as 1882, the physiological psychologist Johannes von Kries expressed the common feeling that attempting to measure subjective feelings was a mistaken attempt to emulate physics, and sixty years later, in 1940, the physicist and philosopher of science Norman Campbell felt able to claim that numerical measurements of subjective experiences were not possible.

Given this scepticism, it is perhaps not surprising that the earliest success stories in psychological measurement occurred in the realm of psychophysics—the area most closely linked to the physical sciences. It was only later that formal methods for measuring strength of attitudes, opinions, preferences, and aspects of personality were developed.

In some ways much psychological measurement can be more difficult than measurement in the natural sciences: inanimate objects cannot observe you trying to measure them and consciously decide to act in a contrary way. So-called 'gaming', where people give the responses they believe you want or from which they derive most benefit, and feedback loops, are common in psychological measurement situations. This is particularly the case when sensitive topics are being studied (e.g. financial, medical, or sexual matters). Sophisticated 'randomized response' methods have been developed to cope with such matters. Similar challenges also arise in social measurements (which are, after all, in some sense aggregates of individual behavioural measurements). The *Hawthorne effect* describes how the mere fact that a person is taking part in a study can influence the responses they give. So, for example, in studies of how working conditions influence industrial productivity, having psychologists going around asking questions—and apparently showing an interest in the workers' conditions—can lead to improved productivity.

The very difficulty of psychological measurement has meant that it has been the focus of a vast amount of research attention. Unfortunately, not everyone appreciates the need for such care when undertaking measurement in the behavioural sciences. As psychologists Robert Hogan and Robert Nicholson observed: 'the literature is replete with examples of researchers testing substantive hypotheses with homemade and unvalidated scales; when it is later discovered that the scales did not measure what they purported to measure, the entire line of research is called into question.' And Henrica de Vet and her colleagues have pointed out: 'Developing a measurement instrument is not something to be done on a rainy Sunday afternoon. If it is done properly, it may take years.' Of course, the fact that some people use measuring instruments carelessly or improperly makes psychology no different from any other discipline.

As with medicine, we should recognize that there are different high level purposes for which psychological measurement might be undertaken, and that these purposes require different kinds of procedures. For example, in educational assessment, we might be testing so that we can identify the areas where students need more instruction, or so that we can tell prospective employers what the strengths and weaknesses of graduates are, or so that we can tell how ready a student is for some activity (e.g. the SAT, assessing readiness for college studies), or even, at a higher level, so that we can see how much value an educational establishment adds to its students. And a similar diversity of aims exists in other domains. Early work on measuring intelligence, for example, was aimed at screening, with Alfred Binet, one of the earliest researchers in the area, developing tests at the start of the 20th century to identify which school pupils would benefit from extra help. For many years such tests were used to stream children into different types of secondary education in the UK. Likewise early tests were used to measure literacy and to determine aptitude for different roles amongst US soldiers in World Wars I and II.

We can also distinguish between measuring to understand something, and measuring for operational purposes—to make a decision, improve a process, and so on. Clearly the former is closely related to representational measurement, and the latter to pragmatic measurement.

Measuring sensation

Physical stimuli elicit a sensation within a person. Thus a listener hears the *volume* of the television in terms of its *loudness* and the *frequency* of a tone as its *pitch*. The *physical amount* of light reflected from this page is perceived as a *degree of brightness*. Chapters 2 and 3 have discussed the measurement of the physical phenomena. Since they occur outside of people, they are susceptible to objective measurement—different people can carry out the same measurement operations and hopefully arrive at the same value. But perceptions and psychological characteristics such as perceived loudness, brightness, bitterness, or sweetness, are intrinsically subjective.

The perceived loudness of a sound is not merely affected by the physical intensity of the stimulus, but is also influenced by other things. It will depend on the mixture of tones, the rate at which the sound ramps up and decays, and also on the context: a whisper in a concert hall might sound much louder than one on a railway train. For brightness, a classic example shows a particular shade of grey, reproduced in two different parts of a picture, but which appears completely different in the two places because of the surrounding shades. Because of such complications, experimenters attempt to strip away the context and remove potentially distorting influences. For example, in measuring loudness, they will focus on a single frequency of sound.

Given pairs of observations on both the physical and perceived magnitude of a stimulus, we can attempt to discover the relationships between the two. Such relationships are called 'psychophysical laws'. A very important example is the Weber–Fechner law, developed during the 19th century. This is based on the observation that the smallest detectable difference between two physical stimuli (the 'just noticeable difference') is often approximately proportional to the magnitude of the underlying stimulus. Put more formally, it says sensation is a *logarithmic* function of physical intensity: $S = \log P$, where S is the perceived magnitude and P is the physical magnitude. This means, for example, that equal changes in the perceived magnitudes of a stimulus occur when the physical magnitude doubles from 1 to 2 to 4 to 8, etc. This proved to be a good working rule, and it held sway for a century (which is pretty good for any area of science).

However, around the middle of the 20th century, S. S. Stevens collected experimental data using a so-called *magnitude estimation* procedure, in which a series of physical stimuli are presented to a respondent who then assigns a number to each, proportional to the perceived magnitude. Stevens's data led him to propose that perceived and physical magnitudes were related by power laws. These are laws of the form $S = aP^b$, where b is a constant which depends on the physical property being studied, and a defines the units of the perceived magnitude. This has turned out to be a good approximation to many real phenomena, including loudness, brightness, odour, temperature, duration, and others. Writing in 1965, Gösta Ekman and Lennart Sjöberg said, 'as an experimental fact, the power law is established beyond any reasonable doubt, possibly more firmly established than anything else in psychology'.

We can see from this that whereas Stevens measures the magnitude of the sensation directly (his participants report the numbers), Fechner *defines* the basic difference between values (that is, he defines the basic unit) as the just noticeable difference, and then builds up the underlying sensation magnitude scale from these differences. However, there is a closer relationship between these two perspectives than at first might appear. In particular, if we take logarithms in Stevens's model we obtain $\log S = \log a + b.\log P$. This is simply a generalization of the Weber–Fechner model $S = \log P$, with $\log S$ replacing S. It begins to look as if the difference between the two approaches is really an arbitrary pragmatic choice of whether to represent the perceived stimuli in raw or log terms. To put it another way, the Weber–Fechner law $S = \log P$ was based on the assumption that the just noticeable differences were equal at all levels of the underlying physical stimuli. If we let these differences have log magnitudes (i.e. replace S by $\log S$), we arrive at a special case of Stevens's law.

Magnitude estimation is just one way of teasing out the relationship between physical and psychological quantities. Another approach is to present the stimuli in pairs and ask the respondent to give the ratio of the apparent magnitudes of the stimuli in each pair. And yet another (for sound) is to arrange for the respondent to adjust a volume control until it appears half as loud as a given volume.

A rather unusual example of a subjective phenomenon is the Scoville scale of the 'hotness' of spicy food. The measurement process begins by extracting the capsinoids (which cause the

5. Carolina Reaper pepper pods, one of the hottest of peppers.

sensation of hotness) and then the heat level is based on the amount of dilution needed until it can be just detected by a panel of five trained tasters. The record for the hottest pepper is regularly broken. The Carolina Reaper (see Figure 5) has an average score of around 1,500,000 Scoville Units, but has been reported to reach 2,200,000 units.

Psychophysics and psychophysiology represent a basic level of measurement of psychological processes. But they are only the beginning. We now need to move further away from physics and physiology towards the higher mental functions. We need to consider how to measure such things as the repulsiveness of a picture, the endearingness of a puppy, and how to measure attitudes and opinions, as well as mental attributes such as intelligence and mental states like depression and wellbeing. It is clear that measuring these poses formidable challenges beyond those of psychophysiological measurement.

In some areas, specialized methods have been developed. For example, (subjective) Bayesian statisticians interpret probability not as an objective property of objects in the world but as a subjective measure of strength of belief that a person has that a statement is true or an event will occur. This strength of belief can be related to the size of bets the person should be prepared to make that particular outcomes will occur. Eliciting the sizes of these bets can thus lead to measurements of probabilities.

Models of psychological measurement

One way of looking at psychological measurement, essentially a representational perspective, is to suppose that there is a true underlying value, which is captured only approximately by the measurement procedure, so that the observed value is a combination of the true value and measurement error. This is the perspective adopted in *classical test theory*. In intelligence testing, for example, we might argue that there are underlying true values of 'reasoning ability', 'comprehension ability', and so on, and that the response to a specific question arises as a specific combination of these true scores perturbed by measurement error.

With this model, we can see that the way to improve the accuracy of measurement is to concentrate efforts on reducing the error components of observed measurements. That requires us to recognize that there are two rather different kinds of measurement error: systematic and random error.

Systematic error is a contribution to error which is persistent and which occurs each time the measurement is taken. A simple physical example of a systematic error arises when bathroom weighing scales are miscalibrated so that they consistently read 1 kg more than they should (meaning that even when nothing was on the scales they would read 1 kg). Psychological examples of systematic errors are acquiescence propensity, in which a respondent has a general predisposition to agree with the questioner, and social desirability, in which respondents tend to answer in a way they feel is socially acceptable. Systematic errors are also illustrated by the effect that question wording has on survey responses: switching the mode of a question from positive to negative ('do you like X?'; 'do you dislike X?') often leads to inconsistent results.

In contrast, random error is a fleeting aberration specific to a particular measurement occasion. This error is likely to take different values at later measurement attempts, with the values occurring randomly according to some distribution.

Systematic error is difficult to reduce, but (as we saw in Chapter 1) we can reduce random error by repeating the measurement multiple times and averaging the results. However, whereas in physics we can repeat a weight measurement simply by putting the object back on the scales and reweighing it, in psychology there is little point in asking the same question again and again: the person being measured is likely to remember their previous response and give it again. Instead, a new and therefore necessarily different question has to be asked. That raises the awkward possibility that differences in the responses to different questions may be due to more than simple random error and may have a systematic component.

Classical test theory measures the effectiveness of a test in terms of its *reliability*, the ratio of the variance of the underlying true score in the population to the variance of the total score. (Total variance is the variance of the true score plus the variance arising from the measurement error.) This is equivalent to simply using the ratio of the error variance to the true score variance: the smaller this ratio, the more accurate is the test, and the greater the reliance that can be placed on individual results. However, since the true score is never actually observed (that's the whole problem!), subtle statistical methods have to be devised to measure this ratio.

Elaborate methods of 'item analysis' have been developed to aid in choosing which questions or items to include in a measuring instrument. For example, it would probably be unwise to include an item which was poorly correlated with the other items, since at best this would be injecting substantial random variation and at worst it could be injecting systematic variation, measuring something not closely related to the aim of the measurement exercise.

The aggregation (averaging or summing) of the scores on multiple different test items, to yield an overall measurement, is an intrinsic aspect of classical test theory. In particular, the scores on the individual items are not of interest. In contrast, in *item response theory* both participants and the items are measured, with the aim being to determine both the difficulty of the test items and the ability of the participants. In this approach, the individual items are binary, having only two possible responses, most generally taken to be 'correct' or 'incorrect'.

We can compare the items in terms of their difficulty by seeing how well people perform on them: a tougher item will elicit fewer correct responses. Or, alternatively, we can compare people in terms of their ability by seeing how many items they answer correctly. However, there is another way to look at this. We can, instead, focus on the interaction between the items and people. After all, the raw observation is not the comparison between people, or the comparison between items, but rather whether or not a particular person gets a particular item right. This leads us to the notion that item difficulty and person ability can be scored on the same scale—with, for example, a person's ability score being greater than (or 'dominating') an item's difficulty score if that person answers that item correctly. This is the principle followed in Guttman scaling, where, as we saw in Chapter 2, ideally a respondent would be able to give correct answers to all questions with a difficulty level below their ability level, and no correct answers to questions with a difficulty level above their ability level. Of course, that is an ideal, and in practice various sources of uncertainty mean that a consistent order may not be achieved.

So-called Rasch modelling supposes that each item has a difficulty value (say, item i has difficulty s_i) and each person has an ability value (say person j has ability s_j), and then the *probability* that person j will give the correct answer to item i is modelled as a function of the difference between the values s_i and s_j: $s_j - s_i$. The greater the difference (so, the more able the person or the easier the item), the more likely it is that the response will be correct. The probabilities of correct responses, estimated by giving a range of people a range of items, can then be used to derive estimates of the item difficulty scores and the person ability scores.

Item response theory is a generalization of this basic Rasch model. For example, in addition to each item having a difficulty score, they may also have a *discriminability* score, showing whether a slight difference in respondent ability is sufficient to yield a dramatic change in the probability of correct response to an item, or if a large difference in respondent ability is needed. The basic Rasch model assumes that similar differences in respondent ability produce similar differences in all items.

An important property of Rasch models is what is known as *specific objectivity*. This means that, for data which fit the model, the difference in ability scores between pairs of people is the same whatever sample of items is used in measuring the difference. In a deep sense this is a requirement for representational measurement: it is equivalent to saying that the score is a real property of the person being measured. Moreover, in a similar way, the differences in the difficulty scores of a pair of items will be the same whatever the sample of people used in conducting the measurement, so the difficulty scores of the items are real properties of the items.

Particular challenges of measuring the mind

The scope and range of measurement in the social and behavioural sciences is illustrated by a 1992 American Psychological Association advertisement in *Monitor*, which estimated that 20,000 psychological measures are created each year. I doubt that the rate of creation of tests has declined since then. In part this diversity reflects the range of different psychological attributes which people want to measure, and in part it represents the challenges of measurement in this domain.

We have already seen that slightly different ways of wording a question can lead to substantially different answers, and that switching between positive and negative formulations can lead to inconsistent responses, but there are many other sources of uncertainty and error. They include ambiguity in the item questions (so they might be interpreted in different ways), difficult questions (which require excessive effort to answer), over-long questionnaires (which exhaust the respondent and lead to inaccurate answers), double negatives, cultural differences, halo effects, leniency effects, range restriction effects, as well as interaction between the tester and testee.

The implication of all this is that considerable care needs to be taken when deciding on precise wording. Normally test development will require multiple extended iterations—recall the comment about test development taking years.

Psychological measurement instruments can also come in a wide variety of types, from the simple and brief self-administered questionnaire, through to the extensive structured interview, where the interviewers have to undergo lengthy training beforehand. Needless to say, instruments of the latter type will tend to give more reliable results—but at a cost. There are also different modes of testing: in Chapter 4 we mentioned visual analogue scales and semantic differential scales, and these and others are also used in psychological measurement.

Self-administration and structured interviewing do not exhaust the range of possible ways of administering psychological tests. Increasingly, tests are administered by computer. This has the merit that, as the test proceeds, so subsequent questions can be chosen in response to those that have gone before, to more precisely focus on the respondent's level of the attribute in question, and so that questions they could easily answer or had no hope of answering could be sidestepped. Computer administration also has the merit of removing biases introduced by interviewer/interviewee interaction. Other tests do not involve interviews or questionnaires at all, but require the participant to complete some exercise, such as a simple physical task, while their performance is observed and rated by someone else.

Many psychological measurement results are given relative to the distribution of scores in some population. This is done with IQ, for example, which is explicitly defined in terms of performance relative to the average of a norm group. But it is also latent in other measures, where, in one way or another, the scores of a population or sample provide a framework with which an individual can be compared. So-called *norm-referencing* does this explicitly. Thus we might give a student's score relative to a distribution of scores. A prospective employer can take reassurance from the fact that the candidate is in the top 5 per cent of students. In contrast, in *criterion-referencing*, the score is given relative to a fixed criterion. For example, we may be told that the student scored 96 per cent on the test. To interpret this, we need some understanding of the difficulty of the test.

An example: measuring intelligence

The measurement of intelligence provides a good illustration of the challenges of measurement in the behavioural sciences.

Although some people ridicule the very notion of measuring intelligence, it is clear that we all think of intelligence as quantitative in some sense: we say that X is very intelligent, or more intelligent than Y. The question is, how can we give more concrete numerical values to this apparently quantitative attribute?

As I have hinted, the obvious starting point is for the participant to answer a series of test items, each tapping into what we think of as intelligence. So, for example, we might have items which test reasoning ability, visuo-spatial perception, verbal ability, and so on. This is what is done in IQ testing, where the scores of a number of test items are added. It is apparent that this is a strongly pragmatic measure: we simply identify relevant items and define the score as the sum of their test values.

Alfred Binet elaborated this basic notion, and came up with the idea of measuring IQ in terms of the age at which an average child would achieve a given level of performance. If a 5-year-old achieved the same level of scores as would be achieved by an average 6-year-old then the 5-year-old's score would be 6/5 = 1.20. Binet then standardized the scores by multiplying them by 100, yielding an IQ of 120 in this example. And here we can see where the term IQ, *intelligence quotient*, comes from: a quotient is a ratio. Later, David Wechsler devised a variant: the ratio of the 5-year-old's score to the average score of 5-year-olds. Still a quotient, so still an IQ. Note that both of these approaches yield scores relative to a given population—we spoke of 'the average score of 5-year-olds', with a specific population of 5-year-olds implied.

Because IQ is based on a sum of distinct items, we should expect it to have an approximately Gaussian distribution. This is purely a consequence of a statistical phenomenon called the *central limit theorem* and it does not tell us anything about the 'distribution' of any underlying intelligence. With this in mind, Wechsler rescaled his scores to have a mean of 100 and a standard deviation of 15, so that half the population had a score between 90 and 110.

A rather different, and much more sophisticated, approach to measuring intelligence owes more (though not everything) to representational measurement. If we observe positive correlations between a set of test items, then we might seek to explain these correlations in terms of the relationships of the test items to a single underlying attribute (or 'factor', and hence *factor analysis*), namely 'intelligence'. In particular, if for each fixed value of the underlying attribute, the test items are uncorrelated, then their overall relationship arises purely because of their relationship to the underlying attribute. The literature on intelligence testing calls this factor *g*, for 'general' intelligence. It means that each item's score is a weighted combination of the underlying factor and a contribution which is unique to the particular test item. By inverting this relationship, it is possible to say what score on the underlying variable, the *g* factor, corresponds to a given configuration of observed test item scores.

Unfortunately, the mathematics of factor analysis shows that the scores on *g* are not uniquely determined by this method. All that is determined is their order—only an ordinal measure results. For this reason, it is usual to apply a pragmatic restriction, and assign scores so that the distribution is Gaussian (with mean of 100 and standard deviation of 15, again). It is important to recognize that this is very different from the Gaussian distribution arising from the way IQ was constructed, even if both turn out to be sums of multiple items. The IQ approach leads to an approximately Gaussian distribution simply because of a statistical property of sums, while the factor analysis approach explicitly models the underlying distribution as Gaussian. The factor analysis approach has the marked advantage that, if no single common factor provides a good explanation for the correlations, we can extend the model by postulating several such underlying factors, each tapping into a different aspect of intelligence. There has been a great deal of research on developing measures of different aspects of intelligence—with different theories producing different numbers of 'types' of intelligence.

IQ may appear particularly simple, being merely a ratio based on sums of item scores. But in applying this approach we must recognize that a simple unweighted sum is equivalent to equally weighting the tests: it is just as much an arbitrary choice as any other. Ideally we need to find an objective way to provide the weighting, and factor analysis provides this.

The difference between the two approaches can be seen if we compare two intelligence tests, each of which has items covering many different kinds of ability, but one of which includes just a handful of items relating to arithmetic ability, while the other includes many. Clearly a measure of IQ based on the second of these two tests will weight arithmetic ability much more heavily, since many arithmetic items will contribute to the overall score. This is not ideal: we need some way of avoiding the fact that the arbitrary choice of the relative numbers of items we use for different aspects of intelligence will influence the final score. Factor analysis does this: the many different arithmetic ability items will be relatively highly correlated, since they are tapping into the same thing, and factor analysis will identify this common aspect.

Of course, if our overall set of items includes none at all relating to arithmetic ability, then we should not expect the final measure to reflect this aspect of intelligence, whether we use IQ or *g*. While sophisticated measurement procedures can do amazing things, they cannot perform miracles.

We see from all this that a major weakness of IQ is the initial choice of items. Change these, and you can change the IQ score. This is not so serious a problem with *g*, where, provided the set of items include all relevant aspects of intelligence, the resulting score will take them into account appropriately. Another way of looking at this is that IQ is a pragmatic reduction of a multidimensional space of different aspects of intelligence to a single dimension, whereas *g* explicitly acknowledges the multidimensional nature of intelligence and measures the most dominant aspect. Of the two, *g* is much to be preferred—but confusion between IQ and *g* explains much of the controversy in measuring intelligence.

Some of the difficulties of measurement in the behavioural sciences are clearly illustrated by this example of measuring intelligence. The bottom line is that there is no etalon, no object with a basic unit size, which can be used to rate an individual, with all this implies about arbitrariness and the need for pragmatic constraints to make the measure unique. Furthermore, if we use different test items to test two people, then we will have doubts about the comparability of the resulting measurements. Similarly, if we developed intelligence measures using different populations, then IQ would be standardized differently in the populations. It is also likely to mean that the weights used in calculating individual *g* measures would differ between the populations, and we would need to satisfy ourselves that they were sufficiently similar if we wanted to claim that the same thing was being measured.

It would be an oversight, in the context of intelligence testing, not to mention the Flynn effect, as an illustration of the particular challenges associated with behavioural and social measurement. The Flynn effect describes the well-documented observation that intelligence test scores plotted over a hundred-year period show a regular upward trend. The steepness of the slope suggests this is not a real effect in intelligence: our grandparents and great-grandparents were not idiots. One proposed explanation is that people are becoming more familiar with the sort of abstract mental manipulations characteristic of intelligence tests as the world comes to rely more and more heavily on symbols and symbol manipulation.

Chapter 6
Measurement in the social sciences, economics, business, and public policy

If, as we noted in Chapter 1, the dawn of measurement lay in the need to determine basic physical concepts such as weight and length, the development of measurement technology owed a great deal to the need to be able to measure aspects of society. Social measurement spans a vast range of topics. It underpins government, public policy, international relations, industrial relations, economics, academic social science research, aspects of business and commerce, and many other areas. It is necessary for understanding our societies and how we live in them, for monitoring and indeed guiding change, to decide if things are working, and to provide accountability. Social measurement is critical in designing our education and health systems, in running our transport systems, and in creating new towns and cities.

We also saw in Chapter 1 that, broadly speaking, social measures are *aggregate* measures, summarizing many individual values. Examples are *median* income, fertility *rate*, GDP, *national* wellbeing, and crime *rate*. This is not the first time we have met aggregate measures, though previously we met them in a rather different guise. In Chapter 5, we saw that agglomerations of scores on different *test items* are used to produce an overall value of the attribute to be measured—such as intelligence or depression. In this chapter, however, our aggregates come about as agglomerations of values of the same attribute from different *objects*, each with its own value. Social measurements are thus statistical summaries of a population.

Statistical summaries might be based on data from every member of a population—from a census, for example (see Figure 6), or from a club or corporation, and we might calculate statistical summaries of the membership (the average age, sex ratio, etc.) from this list. In other situations it might be more convenient (or cheaper, or even more accurate under some circumstances) to base the summary on a mere sample, rather than the entire population. In this case it is clearly critical that the sample is carefully selected so as to avoid biasing the result: a sample based on doorstep interviews taken at 11 a.m. would miss out the substantial portion of the population who were at work at that time. The discipline of survey sampling describes how to draw valid samples and derive estimates based on them.

Some social measures will have a counterpart at the unit level. A country's unemployment rate is the proportion of the population which are unemployed, but we could think of an individual as having an unemployment rate—it would be 0 or 1 according to whether they had a job or not. But other social measures have no unit counterpart. Wealth inequality, for instance, shows how wealth is distributed across populations: it is a statistical summary of the shape of a distribution. And while an individual will have a certain amount of wealth, they will not have their own measure of wealth inequality.

The measurement of aggregate phenomena has progressed via a revealing leapfrog relationship involving the natural sciences and the social sciences. We begin by noting that the word *statistics* has its origin as the study of matters of *state*—that is, of numerical summaries of social and economic phenomena in human societies. Once numbers describing populations had begun to be collected, regularities became apparent—the insurance industry is based on these regularities. George Boole, in his 1854 book *Laws*

Administrative County of London — The undermentioned Houses are situate within the Boundaries of the — Page 31

Civil Parish of Christchurch — Urban Sanitary District of Whitechapel — Parliamentary Borough or Division of Whitechapel — Ecclesiastical Parish or District of Christ Church

No. of Schedule	Road, Street, &c., and No. or Name of House	Name and Surname of each Person	Relation to Head of Family	Condition as to Marriage	Age last Birthday of Males	Age last Birthday of Females	Profession or Occupation	Employer	Employed	Neither Employer nor Employed	Where Born
155	58 & 57	William Baillie	Boarder	M	28		Horse Keeper		X		London Hackney
	Flower & Dean St	Alice do		M		24					Tunbridge Wells
	(Continued)	Charles Murray		M	32		Waiter		X		London St George in East
	Lodging House	Ellen Meek		M		27					Scotland Leith
		Edward Barry		M	25		Dock Labourer		X		London Spitalfields
		William Jones		M	40		Hawker			X	do St Lukes
		James French		M	30		Tailor		X		do do
		Thomas Jenkins		M	36		Bricklayer		X		do Hoxton
		Alice do		M	27						do Chelsea
		David do		S	16						do Whitechapel
		Joseph Davis		M	42		Labourer		X		do Borough
		Henrietta do		M		40					do Lambeth
		Clara Daley		M		34	Charwoman		X		N. K.
		Sarah Byford		M		37	do		X		Suffolk
		Sarah Maloney		M		40	Button Hole T		X		London Whitechapel
		John Connor		M	29	29	Hawker			X	Greenwich
		Hannah Lewis		M		38	do			X	London City
156	58 & 57	Henry Chitty		M	36		Butcher		X		Ramsgate
	Flower & Dean St	Emma do		M		28					Tunbridge Wells
	Continued	Frank Ray		M	36		Hair Dresser		X		Bury St Edmunds
	Lodging House	Mary Ann do		M		41					Halstead [illegible]
		John Eason		M	36		Hawker			X	London St Georges East
		Ellen do		M		25					Ireland [illegible]
		George Stevens		M	30		Bricklayers Labourer		X		London Shoreditch
		Mary A do		M		24					do Hoxton
		Henry Englebach		[illegible]		67					do Whitechapel
		William Bowser		M	40		Gas Stoker		X		do Bethnal Green
		Betsy do		M		36					do do
		James Potter		S	36		Clock Maker		X		Cambridge
		Kate English		S		27					London St Georges East
		Sarah Lilley		S		20	Hawker			X	do Bethnal Green
2			Total of Males and Females		15	17					

Note.—Draw the pen through such of the words of the headings as are inappropriate.

6. An extract from the 1901 UK census.

of Thought, wrote: '...phænomena, in the production of which large masses of men are concerned, do actually exhibit a very remarkable degree of regularity...'. And in his three-volume work *History of Civilization in England* (published between 1857 and 1878), Thomas Henry Buckle noted the regularity of phenomena as diverse as suicide rates and unaddressed letters.

Regularities open the possibility of mathematical description, suggesting that social phenomena might be described in a way analogous to physical phenomena—the equations of mechanics, electricity, etc. So it is no coincidence that the social statistician Adolphe Quetelet gave his seminal 1835 book *A Treatise on Man: And the Development of his Faculties* the alternative title *Essay on Social Physics*. In fact, he wrote: 'In giving to my work the title of Social Physics, I have had no other aim than to collect, in a uniform order, the phenomena affecting man, nearly as physical science brings together the phenomena appertaining to the material world.'

The particular characteristic of this 'social physics' is that the laws derive from the individual behaviour of a very large number of distinct and largely independent constituent elements, each behaving in different and perhaps apparently random ways. This observation, that the aggregate regularity in social measurement was derived from the unpredictable behaviour of the underlying large numbers, was noted by physicists and became one of the drivers behind the development of statistical physics—of thermodynamics. Indeed, the physicists James Clerk Maxwell and Ludwig Boltzmann both referred to Buckle's book, noting that it was the aggregation that made the regularities in society apparent: Maxwell had read Buckle's work a year before he started his studies on the kinetic theory of gases. Maxwell also knew of Quetelet's use of the 'error law', an early term for the central limit theorem we met in Chapter 5, which says that averages or sums tend to have Gaussian distributions (and which Quetelet in turn had adopted from its earlier use in astronomy). Maxwell suggested that the same mathematical description could be applied to the distribution of the velocities of gas molecules.

But the story does not end there. Researchers in the social sciences, noting the success of statistical physics in modelling natural physical phenomena as large-scale aggregates of tiny constituents, began explicitly to emulate it. For example, the thermodynamic models in the paper *On the Equilibrium of Heterogeneous Substances* by the American physicist Willard Gibbs inspired Nobel Laureate economist Paul Samuelson's book *Foundations of Economic Analysis*, and eminent economists such as Irving Fisher and Jan Tinbergen (winner of the first Nobel Prize in economics) both originally studied physics. More recently still, a subdiscipline of *econophysics* has sprung up, explicitly modelling social and economic affairs using statistical physics approaches.

So we see an alternation, of the social sciences taking ideas from the physical sciences, which then take ideas from the social sciences, which then take ideas from the physical sciences, and so on, driven by the fact that in the physical sciences the macroscopic objects are evident, with the microscopic constituents (atoms and molecules) being conjectured, while in the social sciences the constituent objects are very apparent with the 'reality' of the aggregate being less obvious. The story of the development of these ideas illustrates how measures can lead to new concepts.

It is very apparent that there is an infinite number of aspects of society that could be measured. In education, we could measure pupil/teacher ratio, cost per pupil, rate of persistent absence from school, success rate in apprenticeships, countless measures of teacher and school success rates, and so on. In health we could measure under-75 mortality rate from cancer, number of people quitting smoking, per cent of eligible people receiving free health checks, number of emergency hospital admissions for various causes, countless hospital statistics, and so on. In monitoring the performance of local authorities we could measure net cost per person, population per police officer, crime rates, highway maintenance cost per kilometre, per cent of troubled families meeting various criteria, and so on. The list of potential attributes is truly endless. What this means is that, perhaps even more so than in some other measurement contexts, considerable thought must be given to the objective of the measurement exercise.

One strategy to cope with this multiplicity of potential measures is to use a profile of several measures, trying to span the space of what we consider important. It perhaps goes without saying that the profile should not involve too many measures—there have been cases of organizations using hundreds, which is counterproductive. It has been suggested that ten to twenty is about an optimum number, though this will depend on the application. The Shanghai Academic Ranking of World Universities uses six indicators: the number of alumni winning Nobel Prizes or Fields Medals, the number of staff winning Nobel Prizes or Fields Medals, the number of highly cited researchers, the number of papers published in *Nature* and *Science*, the number of papers indexed in the Science Citation Index and Social Science Index, and a weighted sum of the preceding scores divided by the number of full-time equivalent staff.

Inevitably, once a profile has been created, there will be a tendency to combine the separate measures to yield a single score. In many social contexts, this pressure will come from the media, keen to produce league tables. In other contexts, a single index is necessary to guide decisions. For the Shanghai Academic Ranking of World Universities, a weighted sum is produced, using clearly specified weights. The pragmatic nature of this exercise will be very clear, and other university ranking systems use different indicators and weights.

Incidentally, while mentioning league tables it would be remiss not to warn about overinterpretation. Inaccuracies in the raw data can often lead to substantial uncertainty about the precise position in a league table, so that it is not unusual to find considerable jumps in position from year to year. This caution applies elsewhere also: a tiny change in GDP may be the focus of media attention, but may well be irrelevant given the accuracy with which it is measured.

Similar issues apply in sports and games. In some sports, deciding what attribute to measure to determine the ranking of competitors is relatively straightforward. For example, success in a 100-metre sprint or motor racing will be measured in terms of the time to complete the race. Although the mechanics of the measurement technology may be complicated, there is no doubt that elapsed time is the relevant attribute. In other sports, however, things are more subtle. In the decathlon, for example, with achievement in different events measured using different units (of distance, time, weight, etc.) some way needs to be found to combine them to yield an overall score. Needless to say, this is largely a pragmatic choice.

League tables in sports and games have been based on many different ranking and rating systems. Examples include the Colley method, the Borda count, the Elo method (especially for chess), and the Keener method. While such methods are often related to more formal mathematical approaches, such as the Bradley–Terry model mentioned in Chapter 2, they also often have an ad hoc aspect, reflecting their roots in particular practical traditions.

Much social measurement has complications similar to those of behavioural measurement, and not possessed by measurement in the natural sciences and engineering. In particular, people can manipulate measurement processes, perhaps deliberately trying to mislead. The quantification of fraud is an example. In a study of the amount lost due to fraud in the UK, Gordon Blunt and I commented that difficulties arose from formulating basic definitions of what constituted fraud, from the fact that fraud was deliberately concealed (was only the tip of the iceberg detected?), and from the fact that the nature of fraud changed over time as new methods of detection and prevention were adopted.

Sometimes distortion arises as a side effect of the best of motives, as a consequence of a poorly constructed measurement strategy. If one or a few measures are used to assess a social system, there is the risk that other aspects will be ignored. The classic (fictitious!) anecdote is that of the nail factory, in which performance is measured by the weight of nails produced—with the consequence that the factory produced a single giant, and very heavy, nail. But such situations are not always fictitious: they do also arise in real life. Grade inflation in education is a familiar example. Students will prefer to attend classes where the professors award higher grades. They will also rate these courses and professors more highly in their feedback. Higher marks look better on a transcript to be presented to employers, so universities known to award higher grades might seem more attractive. All of these and other factors conspire to drive grades upwards.

Apart from all those potential problems, to be useful social measurements must not be too volatile, varying erratically over time. But neither must they be too static, not reflecting relevant changes.

Economic indicators

Index numbers are widely used, especially in economics, where examples include stock market indices (e.g. the FTSE100, the Dow Jones, and the Nikkei), labour market indices, output measures (such as GDP, GNI, and GNP), and measures of earnings, productivity, wellbeing, deprivation, and prices. I shall illustrate some of the basic ideas underlying such measures using price indices as an example.

As every reader will know, prices fluctuate, often tending to increase over time. Measuring this so-called *inflation* is important for many reasons—so that pensions can be increased appropriately, so that real rates of return on investments can be calculated, to inform wage negotiations, and so on. By measuring inflation, we measure how the purchasing power of money changes over time.

The exercise would be straightforward if there was just a single 'good' or 'service': we would simply compare its current price with an earlier (baseline) price and see by what percentage it had increased. Unfortunately, however, there are complications.

First, we buy a great many different products and consume many different services, and their prices change at different rates. To obtain an overall measure of inflation, we need to imagine a collection of items—a so-called 'basket' of items—and somehow combine their price changes to give an overall measure of change. Just to give an indication of the scale of such exercises, to construct the Consumer Price Index in the US, 'economic assistants' visit shops, service establishments, and other outlets and record the prices of some 80,000 precisely defined items each month, while data are also collected from about 30,000 weekly diaries and 60,000 quarterly interviews to determine what people are buying. The analogous exercise in the UK involves collecting prices of a basket of around 700 goods and services, in each of about 150 towns and cities across the UK each month, making around 115,000 prices. The extent of these exercises indicates why collecting data automatically over the web can have significant advantages, and we can expect more changes in this direction over the next few years.

Second, technological progress means that some items become less important over time, while others become more important, and even perhaps almost universally owned. While a couple of centuries ago candles might have figured prominently in the weekly shopping basket, they are far less important now: it would be questionable whether including changes in the prices of candles was useful for our overall measure of price inflation in the modern world. At the opposite end of the spectrum, mobile phones were non-existent not so long ago, but are now an important aspect of consumer purchases.

In fact, candles and mobile phones together provide a good example of a third point. There are candles and there are candles. There are scented candles, shaped candles, and a wide variety of other specialist candles, but they clearly play a rather different role in society from that which candles used to play. The nature of the product itself has changed, casting doubts on the wisdom and purpose of a simple comparison of today's price with yesterday's price. Similarly, mobile phones constantly evolve, both in terms of capability and quality. A modern phone may well not be comparable with an older version. More generally, the quality of items may differ, and a cheap poor quality item should not be regarded as equivalent to an expensive high quality version of the same thing.

Even for a particular good, competition between suppliers can well mean that it may not have a single price. It might depend from which store you bought it, or if you bought it in a two-for-one or other special offer.

The items making up the basket are generally reviewed, often annually, to allow for changes such as these. The alternative, while producing a strictly comparable index, would mean it gradually became less and less relevant for the uses to which it was put.

There is another important point which should be made in the context of choosing the items for a basket: different people have different patterns of purchases. People with cars buy petrol, those without do not. Some people smoke, others do not. Retirees may have very different consumption patterns from those in work, or those with young children to support. And so on. All of which means that different people have different rates of inflation. Overall measures such as the Consumer Price Index are aggregate national measures, so that they may represent a very different experience from that of any particular consumer.

There are various different philosophical perspectives on the construction of price indices. *Axiomatic* theory seeks to construct an index which satisfies a number of mathematical properties. *Economic* theory evaluates potential index formulae according to their relationship with economic concepts. The *stochastic* approach postulates that all items have a common underlying rate of inflation, along with their unique random variation about that rate (cf. factor analysis). *Divisia* index theory treats time as a continuous variable, rather than occurring in discrete (e.g. annual) steps.

Examples of the sorts of properties which axiomatic theory might require price indices to satisfy are (i) that the index should remain unchanged if the currency in which all the items are purchased is changed in the same way (e.g. from US dollars to pounds sterling); (ii) the value of the index for time t relative to time s should be the reciprocal of the value of the index for time s relative to time t; (iii) the product of the index for time t relative to time s, and the index for time s relative to time r, should equal the index for time t relative to time r. Various such attractive properties have been proposed, but unfortunately it is not possible to construct an index which satisfies all of them. The consequence is that we have to make a choice about which are considered most important, and this will depend on the proposed use.

All of these potential choices mean that many different ways of constructing indices have been proposed. The Carli index is based on the arithmetic mean, over the items in the basket, of the *price relatives*: a price relative is the ratio of the price of an item at some time t to its price at a base period 0: p_t/p_0. The Dutot index works by taking the ratio of the arithmetic mean item price at time t to the arithmetic mean price at base time 0. The Jevons index takes the geometric average of the price relatives. And, of course, there are others. And we see immediately that we might want to weight the prices or the price relatives in our calculations. Furthermore, although our discussion has thus far been solely in terms of price, there is a matching discussion to be had in terms of quantity, q. The *value* of a good purchased is the product of the price per unit and the number of units purchased, pq, and the total value of a basket of goods will be the sum of these products over the items in a basket. The Laspeyres index at time t is the ratio of these sums for prices at time t and at time 0, using the quantities from the earlier time for both sums. The Paasche index uses quantities at the later time for both sums. The 'earlier' time could be a fixed date, the same for all calculations, or it could change as time progresses. For example, it could be the previous year in each case. This is termed 'chaining', and is common in index numbers produced by national statistics institutes. Chaining means the composition of the index is updated as time progresses.

A great many other indices have been proposed, all with subtly different properties. What will be apparent from this discussion is that the construction of price indices is heavily pragmatic in measurement terms. The axiomatic approach explicitly constructs the measure so that it has certain properties—very much a pragmatic perspective. Even the stochastic approach may best be thought of as a way of conveniently summarizing the diverse price evolutions of a variety of items as time progresses.

This discussion has demonstrated that the construction of price indices is a complex and complicated process, and that there is no single 'right' answer. But other economic and social indicators have further complications. At least price indices are trying to summarize multiple measures of the same kind. In contrast, an *index of deprivation* has to combine diverse dimensions of deprivation, making it explicitly pragmatic. The *English Indices of Deprivation 2010* identifies seven domains of deprivation: income; employment; health and disability; education, skills, and training; barriers to housing and services; crime; and living environment. Several indicators of each of these dimensions are defined, and these are then combined to yield domain scores. This simple description disguises some sophisticated statistical analysis going on behind the scenes. For example, values based on small numbers will be more uncertain, simply because of stochastic variability, so statistical tools called shrinkage estimators are used, and the different dimensions may be standardized and transformed in different ways prior to combining them. Finally, the transformed domain scores are weighted and combined.

National wellbeing provides another illuminating example of social measurement. Classically, the wellbeing and progress of a nation has focused on economic measures, such as GDP. But this has major shortcomings, and is only part of the story. Robert Kennedy expressed this very well in a famous 1968 speech, saying:

> Our gross national product...counts air pollution and cigarette advertising, and ambulances to clear our highways of carnage. It counts special locks for our doors and jails for those who break them. It counts the destruction of our redwoods and the loss of our natural wonder in chaotic sprawl. It counts napalm and the cost of a nuclear warhead, and armoured cars for police who fight riots in our streets...Yet the gross national product does not allow for the health of our children, the quality of their education or the joy of their play. It does not include the beauty of our poetry or the strength of our marriages, the intelligence of our public debate or the integrity of our public officials....it measures everything in short, except that which makes life worthwhile.

A wide variety of national initiatives have been launched to try to produce better measures. These are illustrated by a 2009 report produced by the economists Joseph Stiglitz, Amartya Sen, and Jean-Paul Fitoussi, for the then French President Nicolas Sarkozy. This report made a number of recommendations, including that the household perspective should be emphasized, that the distribution of income, consumption, and wealth (i.e. inequality) should be considered, that non-market activities should be measured, that individual quality of life was relevant, that both objective and subjective measures of individual wellbeing provided key information, that a dashboard of sustainability indicators should be included, and that environmental damage should be taken into account.

We can see from this that while 'national' wellbeing certainly depends on a summary measure of the individual wellbeing of the people within the nation, it also has other very important components which are not measurable by aggregation in the same way. It is also clear that the aspects of national wellbeing are qualitatively diverse, so that combining them into a single measure will be difficult at best—a profile or dashboard may be much more desirable. At less exalted levels than national wellbeing, the notion of a profile or dashboard on measures of performance has become embodied in the phrase 'key performance indicators' or KPIs. The 'key' means they are the important aspects. They are often used in conjunction with targets. Related to this, in management, is the 'balanced scorecard'. These are strategy management tools typically based around a few quantifiable aspects of a business, both financial and non-financial, which can together be used to show the health of the operation and as the basis for control.

Gaming and related issues

Measurement in the areas discussed in this chapter is often associated with competition and comparison. This might be between individuals or organizations, or it might concern the match of performance to a target. The measurement exercise simplifies the comparison, reducing the enormous complexity of the real world to single (or perhaps a few) quantified values, so that we can see if one is larger than another. The trouble is that such a reduction from complex to simple necessarily throws away much of the subtlety of the world being measured. This simplification is one of the things which lies at the heart of many of the objections to measurement and quantification.

The simplification also has another side effect: it may be possible to optimize a single simplified measure in ways other than that intended. The earlier example of the nail factory illustrates this. Another classic example is described by Michael Blastland and Andrew Dilnot in their book *The Tiger that Isn't*. In 2001 the UK government had set the target that ambulances should arrive at category A emergencies within eight minutes of being called. This appeared to result in a dramatic improvement. However, study of the arrival times showed an apparent huge peak in values just within the eight minutes, and a dearth immediately after this. It was clear that the target was being met by distorting the numbers—by playing with the definition of an 'urgent' call.

This sort of *gaming* has been studied in many contexts. Goodhart's law, named after economist Charles Goodhart, says *an economic time series that is targeted becomes distorted and unusable.* Campbell's law says worse: *the more any quantitative social indicator is used for social decision-making, the more subject it will be to corruption pressures and the more apt it will be to distort and corrupt the social processes it is intended to monitor.* Further examples of measures inducing these sorts of effects are hospital waiting times (where a system was introduced whereby arrivals have to wait *before* being added to the official waiting list, so reducing the apparent waiting time), surgeons' success rates (where surgeons improved their rates by accepting only less serious cases), the US Federal National Mortgage Association rewarding executives for reporting higher earnings (even if those later turned out to be misreported), the UK's Research Assessment Exercise (where universities could choose which staff to have rated), and infant mortality rate, which is sometimes used as a measure of the general health of less developed societies, but where focusing on this measure may lead to its reduction without any concomitant improvement in other aspects of society.

Goodhart's law, Campbell's law, and similar observations relate to how a system might be distorted by the choice of measurement used to monitor it. But related issues can arise right at the basic level of data collection. For example, individual respondents might be disinclined to answer sensitive questions truthfully. Randomized response methods, mentioned in Chapter 5, are tools for obtaining estimates of population values of sensitive items, where the respondent might be tempted to refuse to answer or to give false answers: for example, obtaining statistics on the sexual experiences of teenagers, or whether people pay cash to builders for small jobs about the house (avoiding tax). The basic form of such tools is as follows, where we suppose the question has two possible responses, yes and no, with 'yes' being the sensitive response—think of the paying builders cash example. Each respondent is instructed to flip a fair coin, keeping the result concealed, and to answer 'yes' if the coin comes up heads, and to answer truthfully if it comes up tails. The truth for any individual who answers 'yes' thus remains unknown, since the researcher does not know whether the coin came up heads or tails. However, if the true proportion in the sample who pay builders cash is p, then the overall proportion answering 'yes' in the study will be $\frac{1}{2} + \frac{1}{2}p$. Working backwards, we can estimate p from the proportion observed in the sample.

Chapter 7
Measurement and understanding

Accuracy

In Chapter 1, we saw the importance of accuracy in measurement. We noted that navigation requires precise knowledge of position, that engineering requires accurate measurement of size, and that science can require incredibly fine measurements. In science, for example, very slight differences between theoretical and observed values can lead to major discoveries—provided the measurement can be trusted. This word 'trust' also hints at a moral aspect to measurement. Accurate measurements merit respect and confidence. Decisions based on solid evidence carry more weight.

While the quest for increased accuracy may be driven by the needs of applications, the effort to achieve accuracy of measurement itself stimulates other developments, both in terms of measuring instruments and systems, and also in terms of the machinery and methods needed to make more accurate measurements. Length measurement based on dividing a unit length (say a foot ruler) into equal intervals soon runs into limits of accuracy—not least those caused by the thickness of the marks used to define the intervals. This means that the drive for increased accuracy itself can stimulate developments which may lead to advances in understanding and science.

Different kinds of measurements are susceptible to different sources of errors. Measurements involving the direct intervention of humans are vulnerable to fatigue, motivation, carelessness, boredom, distraction, and a host of other causes of distortion. Indeed, different people respond to stimuli differently. Measurements which involve humans writing down numbers can suffer from incorrect recording and transcribing of digits, misplaced decimal points, digit transposition, digit preference (the tendency to round figures to convenient whole numbers), etc. All this is on top of measurement errors arising from instrumental issues, such as floor and ceiling effects (when an instrument's range is restricted).

Electronic and other physical systems will suffer from random variations between measurement replications at the limits of their accuracy, as in the twinkling variation in the brightness of stars due to turbulence in the Earth's atmosphere and thermal noise in electronic circuits. Economic statistics may suffer from missing data (e.g. perhaps not all of the surveyed companies had reported their sales figures). Social statistics will have intrinsic variation arising from the fact that the entire population has not been questioned, but only a sample.

That last example brings us on to one important way to increase measurement accuracy which we met in Chapters 1 and 5. This is to repeat the measurements and take an average. Broadly speaking, an average of several independent measurements will be less variable than each of the individual measurements. 'Independent' here means that the measurement procedure is repeated from scratch—after all, an average of ten measurements in which the last nine were simply copies of the first would have the same variability as just one measurement. But when measurements are repeated from scratch, any random variation in the measurement results might be expected to tend to cancel out: some measurements would be overestimates, while others would be underestimates.

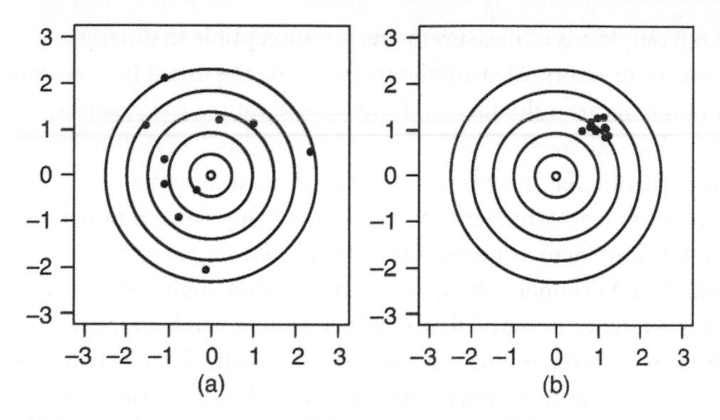

7. (a) Data points are scattered about the centre of the target, but are inaccurate because of their wide dispersion. (b) Data points have a small dispersion, but are inaccurate because their location is not at the centre of the target.

However, as we saw earlier, while averaging multiple measurements will reduce inaccuracies arising from random variation, it will do nothing to improve inaccuracies arising from systematic error: if each of the ten measurements is off by the same amount (as with the miscalibrated bathroom scales in Chapter 5) then no number of repetitions will eliminate the error.

These two fundamental aspects to measurement accuracy are sometimes called precision (how much repeated measurements fluctuate about a central value) and bias (any systematic departure from the underlying true value, affecting all of the repeated measurements). Different disciplines use other words for the same concepts—such as reliability and validity. Figure 7 illustrates these two kinds of inaccuracy.

The precision aspect of measurement—reflecting the way measurement results vary between repetitions—is often reported in terms of error bounds. So, for example, we might see a measurement reported as 5.3 ± 0.5 kg, meaning that the true value is thought to lie between 4.8 and 5.8 kg. Unfortunately, there is no universal standard for defining such intervals. The 0.5 might refer to one standard error, two standard errors, or some other measure of precision—it is necessary to keep alert!

Incidentally, as we saw in Chapter 3, we should not be misled by spurious accuracy reported by large numbers of digits. A claim that the world's population is 7,364,259,981, or that the average weight of men is 81.64663 kg, for cxample, should bc taken with a pinch of salt. The second example illustrates a phenomenon occasionally seen in news reports. The accuracy is totally spurious, and has arisen from a conversion from 180 pounds, which was probably an approximate number to start with. In general, some thought about the measurement procedure which led to the figures can often be revealing.

Measurement and statistics

Here's a straightforward statistical question. I have two boxes, each containing three objects. The objects in the first box have measured values 1, 2, and 6, while the objects in the second box have measured values 3, 4, and 5. Which box has objects with the larger average value?

This is an easy question to answer: the average of the sizes of the objects in the first box is

$$(1+2+6)/3=3$$

and the average of those in the second box is

$$(3+4+5)/3=4$$

so the second box has the larger average value.

For the numbers, this is right, and brooks no contradiction. However, when we ask questions involving measurements, we are not really interested in the numbers per se, but rather in what those numbers tell us about the world.

So now suppose that the objects are diamonds, and the measurements are their weights, in grams. Our calculations showed that the average weight of the diamonds in the second box is larger than that in the first box. Furthermore, since weight is measured on a ratio scale, we could transform the weights by multiplying them by some constant and the resulting values would still represent the empirical relationships between the weights of the diamonds. For example, we could convert the weights of the diamonds from grams into carats by multiplying by five (there are five carats in a gram; the word carat comes from the carob bean, which was once used as a basic unit of gemstone weight). That would give us the values 5, 10, and 30 for the diamonds in the first box, and 15, 20, and 25 for those in the second box, with respective averages of 15 and 20, so that, as we expected, the average weight of the diamonds in the second box is larger than that of the first in carats as well as grams.

But now suppose that, instead, the original measured values represented the hardnesses of two groups of objects, measured on the Mohs scale. As described in Chapter 2, this places a mineral on a scale from 1 to 10, according to whether it can scratch or can be scratched by a set of ten materials of increasing hardness. Then the same calculations would lead us to the conclusion that the average hardness of the objects in the second box was greater than the average hardness of the objects in the first box (still being 4 and 3, respectively).

But, in representational measurement terms, the Mohs scale is merely ordinal. The assigned numbers of 1 to 10 are an arbitrary choice, provided the order is preserved. This being the case, in representational terms we could equally have used the numbers 1, 3, 5, 6, 7, 17, 18, 19, 20, and 21 in place of the numbers 1, 2, 3, 4, 5, 6, 7, 8, 9, and 10 for hardness of the ten objects defining the Mohs scale. This new set of values would be just as legitimate a representation of the *order* of hardness of the ten minerals defining the Mohs scale. And if we used this new set of values, the average hardness of our two collections of objects would be

$$(1 + 3 + 17) / 3 = 7$$

and

$$(5 + 6 + 7) / 3 = 6$$

respectively. Now the *first* box has the greater average hardness.

Note that the comparison between the two groups is not changed by the transformation if the *median* is used instead of the mean. In the first case we obtain the respective medians of the two groups being 2 and 4 before the transformation and 3 and 6 after the transformation. Although the values have changed, the second group still has the larger median after the transformation. Indeed, however the numbers are transformed, provided the order is preserved, the median of the second group will be larger than the median of the first group.

A less trivial example was described in Chapter 2, where we described replacing ordinal scores by the quantiles of a distribution. We cautioned there that such an exercise was not without its dangers. In general, for pragmatic measurements transformations should be seen as part of the definition of the measurement procedure—as part of the definition of what is being measured—so that scales produced before and after such a transformation might be regarded as capturing different aspects of what is being measured.

From a representational perspective, these sorts of issues, whereby transformations of scales can lead to changes in the conclusions, have led to the assertion that the types of scale, and in particular their permissible transformations, determine what statistical operations are legitimate to apply to a set of numbers. But this is not quite correct. The legitimacy of a statistical operation depends on the context in which it is used, and the question it is intended to answer.

Modelling the world

Representational measurement leads to some very deep and powerful tools for constructing models or theories about how the world works. I will illustrate with the simple case of classical physics.

Most variables involved in classical physics are ratio scale variables—think of length, weight, mass, elapsed time, charge, velocity, and so on. That means that the numerical representations of the physical properties are described in terms of an arbitrarily chosen basic unit of measurement, and that we can change the numbers by rescaling.

Now a basic desideratum of a law of nature (or a model, or theory, etc.) is that its form should not depend on an arbitrary choice. In a sense, that's what 'law' means in science: it's a constant form of relationship between several variables. In particular, this means that when we change the units of measurement of one of the variables in a law involving ratio scale variables, the only change should be a change in the units of measurement of other variables, and in particular that the *form* of the law should remain unchanged.

For example, consider Ohm's Law, which tells us the relationship between the electrical properties of voltage (V), current (I), and resistance (R), $V = IR$. Here V is measured in volts, I is measured in amperes, and R is measured in ohms. If we change the current unit to milliamps (by multiplying the value of I by 1,000), while still measuring R in ohms, then changing the units of V to millivolts means the relationship remains the same. Rescaling transformations—the permissible transformations for these kinds of measures—leave the form of the law unaltered. The law does not, for example, become $V = I^2R$.

This is an example of an idea which has proven immensely powerful in establishing physical laws: *invariance*. In this case it is the invariance of the form of the law to ratio scale transformations of the raw variables, but in other situations it is invariance to other kinds of transformations. Indeed, it is no exaggeration to say that the whole of modern physics is built on these kinds of ideas, from Albert Einstein and inertial frames of reference, to Emmy Noether and conservation laws. Gauge theory in physics derives its name from a proposed invariance to certain transformations.

To illustrate the power of the ideas, imagine we are trying to characterize a physical law relating two variables, x and y, both of which we know are measured on ratio scales. Apart from that, all we suppose we know is that x and y are related by some unknown function f, so that $y = f(x)$. Now rescaling x means multiplying it by some constant, k say. Since this must preserve the form of the function, this results in multiplying y by some constant, c, say, so that $cy = f(kx)$. It is easy to see that functions of the form $f(x) = \alpha x^{\beta}$ satisfy this, where α and β are constants. (It is less easy, but also possible, to show that *only* functions of this form satisfy it.)

Here's a concrete illustration. Consider the depth d that a falling object, accelerating under the force of gravity from a stationary start, will have travelled after a time t. Suppose, however, that we knew no physics, and in particular knew nothing about how the distance travelled and the acceleration were related, *apart from the fact that they are both measured on ratio scales*. Then, from the argument we have just seen, we are able to determine that the relationship has the form $d = \alpha t^{\beta}$, with α and β unknown constants. And this is correct: collecting data by observing falling rocks in fact tells us the true relationship is $d = gt^2$ where g is the acceleration due to gravity.

These sorts of ideas generalize to other measurement scales—to interval and ordinal scales for example—producing different kinds of forms that the laws must satisfy, given the scale types.

Sticking with ratio scales, at the end of Chapter 3 I mentioned six basic ratio-scale physical properties (electric charge, temperature, mass, length, time, and angle). Any formula describing some physical phenomenon must be consistent in each of them. This means that the argument leading to the form $f(x) = \alpha x^{\beta}$ must apply to each of them in any formula, and also that formulae must be consistent in terms of these properties—they cannot have length on one side of the formula and length^2 on the other, for example. These ideas lead to very powerful tools (called *dimensional analysis*) for suggesting forms of physical relationships, and also for checking proposed models.

I should mention that there are some formulae which are not dimensionally consistent. For example, there is a classic formula used in child health which says that '4-year-old children are square'. By this is meant that, for such children, height = weight. But this is only true (and only roughly true at that) if height is measured in inches and weight in pounds. It is not invariant to changes of units and cannot therefore be a real physical relationship. However, and this is really the point of these dimensionally inconsistent formulae, it can be a useful heuristic guide, in this case to whether a child is thriving (or, at least it used to be, in those countries where height and weight were measured in Imperial units).

Conclusion

As I have stressed repeatedly throughout this book, measurement provides a window through which we can view the world. We map from the complexity out there to simplified models, defined in terms of measurements. And then we can investigate, predict, explore, understand, and control the world through our models. This is true for trade and commerce, for government and politics, for medicine and science, and for all other aspects of life. It is even true for sports, where the precise measurement of performance has led to incremental improvements in a wide range of activities, including baseball, football, cycling, and athletics.

At several points in the narrative, I have remarked that people have often been suspicious or even worse about efforts to expand the reach of notions of measurement. This seems to have been the case throughout human history. Early medical researchers resisted measuring something so evidently quantitative (to us, anyway) as heart rate. Even that arch quantifier Adolphe Quetelet drew a line, remarking: 'How can we ever maintain, without absurdity, that the courage of one man is to that of another as five is to six, for example, almost as we should speak of their stature? Should we not laugh at the pretension of a geometrician, who seriously maintained that he had calculated that the genius of Homer is to that of Virgil as three to two?'

Richard Shryock explains it:

> one detects the feeling that measurement somehow robs human phenomena of all mystery or beauty, and denies to investigators the satisfaction of age-old sense impressions and of intuitive understanding. Such feelings usually appear within any discipline when it is first threatened, as it were, by quantification. Dr Stevens terms it, in relation to current psychology, 'the nostalgic pain of a romantic yearning to remain securely inscrutable.'

Gradually, however, each such objection has fallen to the inexorable march of measurement. The defeats have not come about through careful arguments convincing the doubters, but simply because of what could be achieved through measurement.

Theodore Porter sums this up: 'it is important to add that there is no fixed limit to what can be quantified, and that a richly nuanced or profound analysis of a large question is never logically excluded by the attempt to quantify parts of it'.

I opened this book with the words of the 16th-century physician and mathematician Robert Recorde (the inventor of the equals sign). I shall end it with the words of Theodore Porter, describing the work of the mathematician and biostatistician Karl Pearson: 'Karl Pearson was neither the first nor the last to worship quantification, which he regarded as integral to scientific method. Its appeal has been the appeal of impersonality, discipline, and rules. Out of such materials, science has fashioned a world.'

References

Chapter 1: A brief history

Alder, K. (2002) *The Measure of All Things: The Seven-Year Odyssey that Transformed the World*. London: Little, Brown (p. 342).

Alexander, J. H. (1850) *Universal Dictionary of Weights and Measures*. Baltimore: William Minifie and Co.

Condorcet, Nicolas de (1793) *Observations sur le 29ième livre de l'Esprit des lois*, in *Oeuvres*. Paris: Didot, 1847 (pp. 376–81).

Harrington, R. (1804) *The Death-Warrant of the French Theory of Chemistry*. London (p. 217).

Koebel, Jacob (1570) *Geometrei von künstlichem Feldmessen*. <http://reader.digitale-sammlungen.de/de/fs1/object/display/bsb11110899_00010.html> (accessed 6 February 2016).

Laming, D. (2002) Review of 'Measurement in Psychology: A Critical History of a Methodological Concept'. *Quarterly Journal of Experimental Psychology A*, 55: 689–92.

Montesquieu, Charles de Secondat, baron de (1721) *Lettres persanes*. Trans. John Davidson. London: George Routledge and Sons (Letter CXII).

Thomson, W. (1891) *Popular Lectures and Addresses*. London: Macmillan (vol. 1 pp. 80–1).

UNICEF (2013) *Every Child's BirthRight: Inequities and Trends in Birth Registration*, United Nations Children's Fund, New York. <http://www.unicef.org/mena/MENA-Birth_Registration_report_low_res-01.pdf>.

Young, A. (1794) *Travels During the Years 1787, 1788, and 1789*. 2nd edn, London (vol. 1 pp. 315–16).

Chapter 2: What is measurement?

Alder, K. (2002) *The Measure of All Things: The Seven-Year Odyssey that Transformed the World*. London: Little, Brown (p. 342).

Apgar V. (1953) 'A Proposal for a New Method of Evaluation of the Newborn Infant'. *Current Researches in Anesthesia and Analgesia* (July–August): 260.

Bridgman, P. W. (1927) *The Logic of Modern Physics*. New York: Macmillan.

Fayers, P. M. and Hand, D. J. (2002) 'Causal Variables, Indicator Variables, and Measurement Scales, with Discussion'. *Journal of the Royal Statistical Society, Series A*, 165: 233–61.

Gould, S. J. (1996) *The Mismeasure of Man*. London: Penguin Books.

Mill, J. S. (ed.) (1869) *Analysis of the Phenomena of the Human Mind, Volume II*. London: Longmans, Green, Reader, and Dyer (footnote to ch. XIV).

Chapter 4: Measurement in the life sciences, medicine, and health

Fayers, P. and Machin, D. (2000) *Quality of Life: Assessment, Analysis, and Interpretation*. Chichester: Wiley.

NYHA (1994) The Criteria Committee of the New York Heart Association. *Nomenclature and Criteria for Diagnosis of Diseases of the Heart and Great Vessels*. 9th edn. Boston: Little, Brown & Co. (pp. 253–6).

Rehabilitation Institute of Chicago (2010) *Rehab Measures: Sickness Impact Profile*. <http://www.rehabmeasures.org/Lists/RehabMeasures/PrintView.aspx?ID=955> (accessed 28 May 2015).

Chapter 5: Measurement in the behavioural sciences

de Vet H. C. W., Terwee, C. B., Mokkink, K. B., and Knol, D. L. (2011) *Measurement in Medicine*. Cambridge: Cambridge University Press.

Ekman, G. and Sjöberg, L. (1965) 'Scaling'. *Annual Review of Psychology*, 16: 451–74.

Hogan, R. and Nicholson, R. A. (1988) 'The Meaning of Personality Test Scores'. *American Psychologist*, 43: 621–6.

von Kries, J. (1882) 'Über die Messung intensiver Grössen und über das sogenannte psycholophysische Gesetz'. *Vierteljahrsschrift für Wissenschaftliche Philosophie*, 6: 257–94.

Chapter 6: Measurement in the social sciences, economics, business, and public policy

Blastland, M. and Dilnot, A. (2007) *The Tiger That Isn't*. London: Profile Books.

Boole, G. (1854) *An Investigation of the Laws of Thought on Which are Founded the Mathematical Theories of Logic and Probabilities*. Project Gutenberg, 2005.

Hand, D. J. and Blunt, G. (2009) 'Estimating the Iceberg: How Much Fraud is there in the UK?' *Journal of Financial Transformation*, 25/1: 19–29.

Kennedy, R. (1968) University of Kansas address, 18 March 1968. <http://www.youtube.com/watch?v=z7-G3PC_868> (accessed 15 August 2015).

Quetelet, M. A. (1842) *A Treatise on Man: And the Development of his Faculties*. Edinburgh. (Originally published in French, 1835, as *Sur l'homme et le développement de ses facultés, ou Essai de physique sociale*).

Stiglitz, J. E., Sen, S., and Fitoussi, J. -P. (2010) *Report of the Commission on the Measurement of Economic Performance and Social Progress* at <http://www.stat.si/doc/drzstat/Stiglitz%20 report.pdf> (accessed 15 August 2015).

Chapter 7: Measurement and understanding

Porter, T. M. (1995) *Trust in Numbers: The Pursuit of Objectivity in Science and Public Life*. Princeton: Princeton University Press.

Quetelet, M. A. (1842) *A Treatise on Man: And the Development of his Faculties*. Edinburgh. (Originally published in French, 1835, as *Sur l'homme et le développement de ses facultés, ou Essai de physique sociale*).

Shryock, R. H. (1961) 'The History of Quantification in Medical Science'. In H. Woolf (ed.), *Quantification: A History of the Meaning of Measurement in the Natural and Social Sciences*. Indianapolis: Bobbs Merrill (pp. 85–107).

Further reading

This section lists some general reading material going into greater depth on the topics of each chapter.

Chapter 1: A brief history

Klein, H. A. (1974) *The Science of Measurement: A Historical Survey*. New York: Dover Publications.

Kula, W. (1986) *Measures and Men*. Princeton: Princeton University Press.

Chapter 2: What is measurement?

Hand, D. J. (2004) *Measurement Theory and Practice: The World through Quantification*. Chichester: Wiley.

Roberts, F. S. (1979) *Measurement Theory, with Applications to Decisionmaking, Utility, and the Social Sciences*. Reading, Mass.: Addison-Wesley.

Chapter 3: Measurement in the physical sciences and engineering

Chang, H. (2004) *Inventing Temperature: Measurement and Scientific Progress*. Oxford: Oxford University Press.

de Grijs, R. (2011) *An Introduction to Distance Measurement in Astronomy*. Chichester: Wiley.

Keithley, J. F. (1999) *The Story of Electrical and Magnetic Measurements: From 500 BC to the 1940s*. New York: IEEE Press.

Chapter 4: Measurement in the life sciences, medicine, and health

de Vet, H. C. W., Terwee, C. B., Mokkink, K. B., and Knol, D. L. (2011) *Measurement in Medicine*. Cambridge: Cambridge University Press.

Feinstein, A. R. (1987) *Clinimetrics*. New Haven: Yale University Press.

McDowell, I. and Newell, C. (1996) *Measuring Health: A Guide to Rating Scales and Questionnaires*. Oxford: Oxford University Press.

Chapter 5: Measurement in the behavioural sciences

Allen, M. J. and Yen, W. M. (1979) *Introduction to Measurement Theory*. Monterey, Calif.: Brooks/Cole Publishing Company.

Bartholomew, D. J. (2004) *Measuring Intelligence: Facts and Fallacies*. Cambridge: Cambridge University Press.

Laming, D. (1997) *The Measurement of Sensation*. Oxford: Oxford University Press.

Chapter 6: Measurement in the social sciences, economics, business, and public policy

Allin, P. and Hand, D. J. (2014) *The Wellbeing of Nations: Meaning, Motive, and Measurement*. Chichester: John Wiley and Sons.

Bartholomew, D. J. (ed.) (2006) *Measurement*. Los Angeles: Sage Publications.

Temple, P. (2003) *First Steps in Economic Indicators*. Boston: Prentice-Hall.

Chapter 7: Measurement and understanding

Langville, A. N. and Meyer, C. D. (2012) *Who's #1? The Science of Rating and Ranking*. Princeton: Princeton University Press.

Palmer, A. C. (2008) *Dimensional Analysis and Intelligent Experimentation*. Singapore: World Scientific.

Index

D

E

F

G

H

I

J

K

L

M

N

O

P

Q

R

SOCIAL MEDIA
Very Short Introduction

Join our community

www.oup.com/vsi

- Join us online at the official Very Short Introductions **Facebook** page.
- Access the thoughts and musings of our authors with our online **blog**.
- Sign up for our monthly **e-newsletter** to receive information on all new titles publishing that month.
- Browse the full range of Very Short Introductions online.
- Read **extracts** from the Introductions for free.
- If you are a teacher or lecturer you can order inspection copies quickly and simply via our website.